Hepatic Circulation

Physiology and Pathophysiology

Colloquium Series on Integrated Systems Physiology: From Molecule to Function

Editors

D. Neil Granger, ***Louisiana State University Health Sciences Center***
Joey Granger, ***University of Mississippi Medical Center***

This Series focuses on combining the full spectrum of tissue function from the molecular to the cellular level. Topics are treated in full recognition that molecular reductionism and integrated functional overviews need to be combined for a full view of the mechanisms and biomedical implications.

Forthcoming Titles:

Capillary Fluid Exchange
Ronald Korthuis
University of Missoui, Columbia

Endothelin and Cardiovascular Regulation
David Webb
University of Edinburgh, Queen's Medical Institute

Homeostasis and the Vascular Wall
Rolando Rumbaut
Baylor College of Medicine

Inflammation and Circulation
D. Neil Granger
Louisiana State University Health Sciences Center

Lymphatics
David Zawieja
Texas A&M University

Pulmonary Circulation
Mary Townsley
University of South Alabama

Regulation of Arterial Pressure
Joey Granger
University of Mississippi

Regulation of Tissue Oxygenation
Roland Pittman
Virginia Commonwealth University

Regulation of Vascular Smooth Muscle Function
Raouf Khalil
Harvard University

Vascular Biology of the Placenta
Yuping Wang
Louisiana State University

Hepatic Circulation: Physiology and Pathophysiology
W. Wayne Lautt
www.morganclaypool.com

ISBN: 9781615040094 paperback

ISBN: 9781615040100 ebook

DOI: 10.4199/C00004ED1V01Y200910ISP001

A Publication in the Morgan & Claypool Life Sciences series

COLLOQUIUM SERIES ON INTEGRATED SYSTEMS PHYSIOLOGY: FROM MOLECULE TO FUNCTION

Book #1

Series Editors: D. Neil Granger, LSU Health Sciences Center and Joey Granger, University of Mississippi School of Medicine

Series ISSN Pending

Hepatic Circulation
Physiology and Pathophysiology

W. Wayne Lautt
University of Manitoba

COLLOQUIUM SERIES ON INTEGRATED SYSTEMS PHYSIOLOGY: FROM MOLECULE TO FUNCTION #1

MORGAN & CLAYPOOL LIFE SCIENCES

ABSTRACT

The Hepatic circulation is unique among vascular beds. The most obvious unique features include the dual vascular supply; the mechanism of intrinsic regulation of the hepatic artery (the hepatic arterial buffer response); the fact that portal blood flow, supplying two thirds of liver blood flow, is not controlled directly by the liver; the fact that 20% of the cardiac output rushes through the most vascularized organ in the body, driven by a pressure gradient of only a few millimeters of mercury; the extremely distensible capacitance and venous resistance sites; the unidirectional acinar blood flow that regulates parenchymal cell metabolic specialization; and the high concentration of macrophagic (Kupffer) cells filtering the blood. The liver is the only organ reported to have regional blood flow monitored by the autonomic nervous system. This mechanism, when dysfunctional, accounts for the hepatorenal syndrome and offers a mechanistic therapeutic target to treat this syndrome. The trigger for liver regeneration is dependent on hepatic hemodynamics so that chronic liver blood flow regulates liver cell mass. In severe liver disease, the whole body circulation is reorganized, by forming portacaval shunts, to accommodate the increased intrahepatic venous resistance. These shunts protect the venous drainage of the splanchnic organs but lead to loss of major regulatory roles of the liver. The development of knowledge of the hepatic vasculature is presented from a historical perspective with modern concepts summarized based on the perspective of the author's four decades of devotion to this most marvelous of organs.

KEYWORDS

hepatic artery, portal vein, hepatorenal syndrome, liver regeneration, hepatic arterial buffer response

Acknowledgements

My first acknowledgement is to the giants upon whose shoulders I have stood to gain some small vision. Chief among those is Clive Victor Greenway (1937–2008) to whom this monograph is dedicated. Clive introduced me to the concepts of peripheral vascular regulation and served as a superior mentor from my first days as a graduate student in 1969, continuing until his retirement in 1996. I am also grateful to have had the opportunity to serve as a mentor to a young research technologist, Dallas Joseph Légaré, who first got involved (1981) with the hepatic circulation by carrying out pivotal studies related to the discovery of the hepatic arterial buffer response. In his first year, we hosted an international symposium on "Hepatic Circulation in Health and Disease" and published the proceedings. His project management skills were apparent in the organization of the meeting and publication of the book. His surgical and conceptual skills and dedication to quality project management allowed many of our studies to succeed over the next nearly three decades. Many trainees at many levels participated in our research adventures. Collaborators whose science played particularly crucial roles in the material discussed in this monograph include Drs. Helen Wang and Jodi Schoen Smith for the liver regeneration studies, Dr. Paula Macedo for the nitric oxide/shear stress studies (and ongoing diabetes collaboration), and Dr. Zhi Ming for his hepatorenal syndrome studies (and ongoing diabetes collaboration). I would also like to acknowledge the funding that I have had the privilege of receiving for support of these studies over four decades, provided by the Canadian Institutes of Health Research (formerly the Medical Research Council of Canada), the Saskatchewan and Manitoba Heart and Stroke Foundations, the Canadian Diabetes Association, the Saskatoon and Manitoba Health Research Councils, National Sciences and Engineering Research Council, and the Canadian Liver Foundation.

I could not have assembled this monograph from hundreds of pages of notes and scribbles without the capable assistance of Ms. Karen Sanders who has carried out this role since 1984.

I am particularly grateful to my wife, Melanie, who has always understood the concepts guiding my science and has often been my most penetrating inquisitor but always my most devoted supporter. My daughter, Kelly, followed the science as a young child and actually provided the mathematical solution for the index of contractility at the age of 13 years.

Many excellent references have been missed, sometimes because of ignorance, sometimes because of an attempt to provide only a conceptual overview, sometimes to preserve the spirit of a textbook, as opposed to a detailed review of all published works. Wherever controversies are identifiable, I have taken a position using my best judgment. Reference to review articles does a disservice to the original authors and, although reviews are a useful resource, I apologize that it has caused me to neglect citing many original works. I am, however, grateful to the many scholars of the hepatic vascular bed whose work and insights have plumbed the depth and breadth of this unique vascular bed. I hope the offering of this monograph is received in the good spirit with which I wrote it and that its many flaws will be forgiven, and useful knowledge will be extracted by each reader.

Contents

CHAPTER 1

Historical Perspectives

In science, the development of knowledge is largely dependent on the technological status of the science, which determines the chronology of discoveries. Cardiovascular functions have been the subject of speculation since the first caveman consciously considered the significance of blood loss. The description of the circulation of the blood by William Harvey in 1628 allowed, for the first time, the development of accurate knowledge about the hepatic circulation. Glisson, for whom the hepatic capsule was named in the mid-1600s, advanced the knowledge of the gross anatomy of the hepatic vascular bed and demonstrated that portal blood flowed through the liver. Wepfer in 1664 was apparently the first to notice the glandular appearance of hepatic acini beneath Glisson's capsule. Twenty-one years later, in 1685, Malpighi confirmed the existence of similar microvascular units that he redefined as hexagonal lobules. To this day, there is no complete consensus on whether the microvascular unit of the liver should be referred to as a lobule, centering on a hepatic vein, or an acinus, centering on a "portal triad" consisting of a terminal branch of the hepatic artery, portal vein, and bile duct, encased within a limiting plate of cells defining the space of Mall. Although anatomical examinations were pursued, function remained largely unknown. In 1890, the space of Disse was identified and named, but the significance of this narrow space between the liver cell and the vascular endothelial cells was unknown. Claude Bernard and Ernest Starling, in the late 19th century, established the liver as an organ of major endocrine, metabolic, and vascular importance.

The development of microscopy was aided by transillumination of the liver, which allowed for greater visibility of the blood flow and structures immediately beneath Glisson's capsule. The advent of the motion picture allowed investigation in the fourth dimension—time.

In 1954 Child [52] enthused "Today a new chapter in hepatic physiology is being written, not through the study of fixed and stained sections of the liver, not through indirect evidence based upon gross physiological observations, but upon actual observation of living tissue. From the dynamic picture presented by the transillumination technique, Knisely, Mann, Seneviratne, and others have been able to verify many of the complex vascular mechanisms postulated by the microscopic anatomists of the past. For instance, Knisely has identified an inlet sphincter at the junction of the portal venule and the portal vein. Furthermore, he has seen an outlet sphincter located at the junction of the sinusoid with the central vein."

Rappaport [308], describing historical development and the impact of the motion picture recording of transilluminated hepatic vascular beds enthused, "though only morphological information was initially gathered, it enabled a continuous record of events in space and time that one could study and restudy until all information on the living vessels and their ever moving content was extracted."

Rappaport became a strong advocate of the acinus as the functional unit of the liver. The unique microcirculatory anatomy of the acinus allows for blood to flow from the center of the acinus and exit into hepatic venules after passing only 16–20 hepatocytes. He also showed movies of red blood cells spinning around clusters of hepatocytes for several rotations before racing out into the terminal hepatic venules.

Whereas the transilluminated moving view of the hepatic circulation provided views of hemodynamic interactions on the surface of the liver, the advent of scanning electron microscopy of vascular casts allowed visualization of static detailed three-dimensional vascular structures with a great depth of field and access to the entire liver mass [124]. The authors acknowledged that all of the existing connections had previously been demonstrated by other techniques; however, the richness of the image of the three-dimensional SEM photographs added depth to the knowledge. Casts produced by infusing latex through the hepatic artery or portal vein show tufts of sinusoids grouped around the terminal conducting vessels, appearing like crowns of flowers on vascular stalks (Figure 1.1).

In the 1960s, a group of Swedish scientists led by Bjorn Folkow revolutionized the concept of the functions and methods to study peripheral vascular beds by separately considering resistance, capacitance, and fluid exchange functions. They directly applied the principles identified by Starling in 1896 who demonstrated that raising the venous pressure in different vascular beds resulted in lymph formation of dramatically different characteristics, that coming from the liver being high in volume and of a protein content similar to plasma. The counteracting "Starling forces" of hydrostatic pressure and colloid osmotic pressure gradients acting across the endothelial vascular wall explained extracellular fluid exchange dynamics. These concepts were pivotal for the approach developed by the Swedes.

In 1960, Mellander, a senior scientist of the Swedish group, described a technique for the simultaneous measurement of resistance, capacitance, and fluid exchange responses in the vascular bed of skeletal muscle. A plethysmograph containing the intact vascular bed, for example, the hind-limb, was key to the studies. This technique was later extended to the intestinal vascular bed [81] and the spleen [118]. These early conceptual breakthroughs were reviewed by Mellander and Johansson [264].

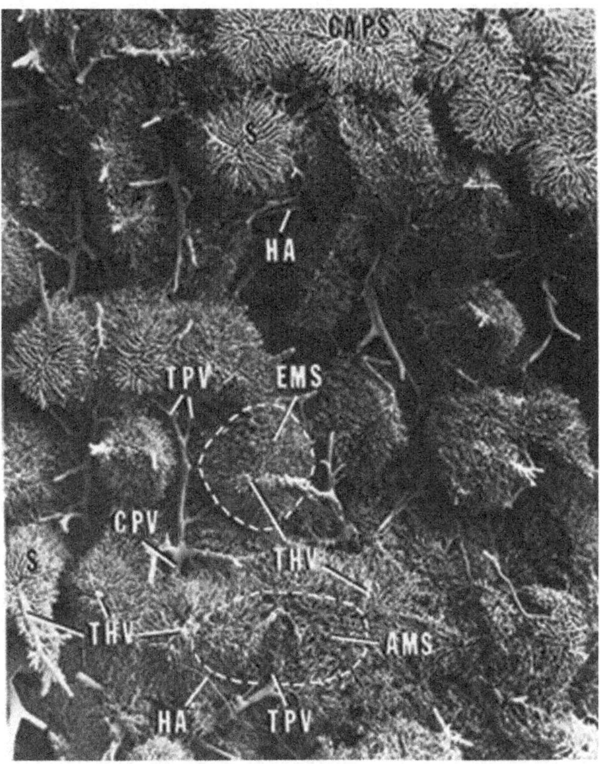

FIGURE 1.1: A cast of the hepatic microvessels of the rat made by perfusing casting medium through the aorta until all hepatic vessels were filled. Both subcapsular sinusoids (CAPS) and deeper vessels are focused. Tufts of sinusoids surround terminal limbs of afferent and efferent vessels. Terminal hepatic veins (THV) collect sinusoids (S), each tuft comprising an efferent microvascular segment (EMS) or hexagonal lobule. Terminal portal veins (TPV) supply sinusoids, each tuft comprising an afferent microvascular segment (AMS) or acinus. Conducting (CPV) and distributing (DPV) portal veins and accompanying hepatic arteries (HA) are shown (magnification, approximately ×40). Reproduced from Grisham JW, Nopanitaya, W. Scanning electron microscopy of casts of hepatic microvessels: review of methods and results. In: *Hepatic Circulation in Health and Disease*. New York: Raven Press, Figure 1, p. 94, 1981. (Figure 1 from publication Hepatic Circulation in Health and Disease is reproduced with permission of publisher Raven Press).

The first use of a crude plethysmograph to measure changes in hepatic volume in vivo was carried out by Francois-Franck and Hallion [83] in 1896 who demonstrated a decrease in liver volume after hemorrhage. Greenway adapted the plethysmograph models of the Swedes to encase the liver in a fluid-filled space without interfering with the vascular inlet or outlet, bile or lymph flow, or hepatic innervation. In addition to assembling a three-piece Plexiglas plethysmograph around the liver and plugging the outlet with an inert gel, plastibase, the research team of Clive Greenway, his postdoctoral fellow Ron Stark, and myself as a graduate student added vascular circuitry to the plethysmograph preparation. This preparation has been fully described [103].

The hepatic venous long-circuit represented a powerful new tool (Chapter 4). The circuit was established by cannulating the inferior vena cava in the thorax to direct lower vena caval blood to an exterior reservoir. The blood was warmed and pumped back to the heart through catheters in the jugular veins. To separate hepatic venous blood from other vena caval blood, the inferior vena cava was ligated proximal to the hepatic venous entrance to the vena cava. Blood from the vena cava below the occlusion was drained in a retrograde manner through venous catheters in the femoral veins, which emptied the blood into the same extracorporeal reservoir. Finally, an electromagnetic flow probe was placed on the hepatic artery. Although this preparation was surgically very complex, the information it provided was unprecedented. Hepatic venous pressure could be accurately monitored and manipulated. Hepatic venous pressure could be increased by elevating the hepatic venous outflow catheter so that pressure, volume, and fluid exchange studies could be readily carried out. Pure mixed hepatic venous blood samples were available for chemical or blood gas sampling. Total hepatic blood flow could be accurately quantified and calibrated, simply by timing the outflow into a calibrated cylinder. Portal venous flow was calculated by subtracting the hepatic arterial flow from total hepatic blood flow. The nerves could be electrically stimulated in this preparation and catheters in the portal vein and hepatic arterial branch could determine blood pressures and administer drugs.

Although in 1896, Starling made initial forays into studies of fluid exchange in the liver, quantification of fluid exchange awaited the plethysmograph just as quantification of solute exchange across intrahepatic vascular compartments awaited the use of multiple tracer dilution technology as carried out by Goresky [97]. Individual tracer compounds or a cocktail of compounds were injected into the portal vein and the efflux of the tracer from a continuously sampled hepatic venous outflow provided novel and readily quantifiable information regarding the passage of substances from the blood into the different hepatic spaces. Red blood cells produced sharp outflow curves as the blood went directly through the liver. Albumin showed greater access to hepatic space followed by sucrose, sodium, and heavy water. This technique gained use for several years to explore normal and diseased livers. Many of the research tools described made a large impact on advancement of knowledge and then, for a variety of reasons, have become valued for the historical development of knowledge and are no longer used or required.

I recommend two books in particular for a historical perspective. The first is a classic, *The Hepatic Circulation and Portal Hypertension* by Child [52]. Child heavily references historical data and reports extensively on morphology. The knowledge of the hepatic and portal circulation in health and disease up to the mid-1900s was detailed in a most scholarly manner.

The second book is one that I had the privilege of editing in 1981. The book was a collection from authors who met at a symposium on *Hepatic Circulation in Health and Disease* in Saskatoon, Saskatchewan, Canada, in 1980. The authors presented orally and contributed chapters covering a comprehensive range of hepatic vasculature knowledge. The full transcripts of the lively discussions add spice. By that time, most of the modern techniques required to evaluate and quantitate hepatic vascular function had been developed. Full knowledge, however, had not yet been extracted.

• • • •

CHAPTER 2

Overview

The liver receives 25% of the cardiac output, although it constitutes only 2.5% of body weight. The hepatic parenchymal cells are the most richly perfused of any of the organs, and each parenchymal cell on the average is in contact with perfusate on two sides of the cell. Of the total hepatic blood flow (100–130 ml/min per 100 g of liver, 30 ml/min per kilogram of body weight), one fifth to one third is supplied by the hepatic artery. About two thirds of the hepatic blood supply is portal venous blood. The gross vascular supply of the liver is conceptually described in Figures 2.1 and 2.2.

The high pressure, well-oxygenated arterial blood mixes completely with the low-pressure, less well-oxygenated, but nutrient-rich, portal venous blood within the hepatic sinusoids. Uptake of compounds by the parenchymal cells and exchange between parenchymal cells and plasma are affected by several unique characteristics of the hepatic microvascular circulation. The characteristics of uptake and exchange by the liver (Chapter 3) have major impact on lipoprotein metabolism, endocrine homeostasis, and therapeutic procedures (drug clearance). While the flow rate of blood through the liver is high, the volume of blood contained within the liver is similarly high and plays a central role in the maintenance of cardiovascular homeostasis (Chapter 4). The liver represents a major blood reservoir in the body; it has a crucial role in the response to blood loss or expanded fluid volume and has a recognized role in determining the response to pressor, antihypertensive, and afterload-reducing agents. Intrahepatic and portal venous pressures are regulated primarily by hepatic venous sphincters, and in the basal resting state, portal pressure is insignificantly different from sinusoidal pressure (Chapter 6). The hepatic circulation has multiple interacting factors that attempt to regulate hepatic blood flow as constant as possible in response to acute and chronic conditions (Chapter 16). The hepatic circulation directly influences renal function through a reflex control, with the afferent sensory limb detecting blood flow-dependent changes in intrahepatic adenosine content and the efferent role acting through sympathetic nerves in the kidney (Chapter 13). This mechanism offers a mechanistic explanation of the hepatorenal syndrome and a therapeutic approach for prevention and treatment of fluid retention.

The constant ratio of blood flow to liver cell mass is regulated in part by adjusting blood flow through the hepatic arterial buffer response (Chapter 5). The flow/mass ratio is also regulated powerfully by flow. Liver blood flow determines liver parenchymal cell volume by a mechanism

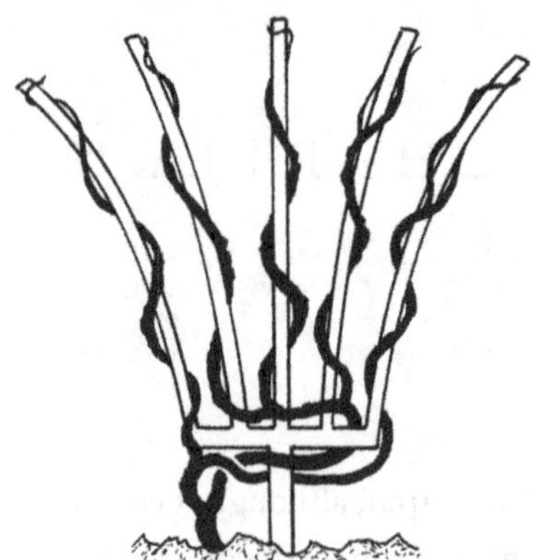

FIGURE 2.1: The branches of the portal vein form a system of a rather constant pattern that is symbolized as a white trellis. The branches of the hepatic artery and the tributaries of the hepatic bile duct become coordinated with the portal venous branches (risers of the trellis) as soon as they are inside the liver. Reproduced from Elias H, Sherrick JC. *Morphology of the Liver.* Academic Press, New York, New York. Figure VIII-11, p. 289, 1969. (Figure VIII-11 from publication Morphology of the Liver is reproduced with permission of publisher Academic Press).

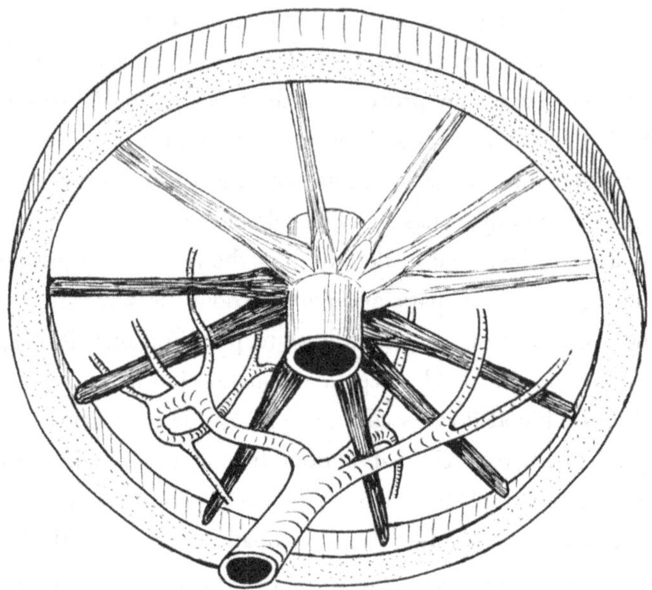

FIGURE 2.2: The hepatic veins (dark), like the spokes of a wheel, are radially arranged around an axle (the inferior vena cava). The portal venous branches wind between them. Reproduced from Elias H, Sherrick JC. *Morphology of the Liver.* Academic Press, New York, New York. Figure VIII-10, p. 289, 1969. (Figure VIII-10 from publication Morphology of the Liver is reproduced with permission of publisher Academic Press).

that is based on the effect of hepatic blood flow to generate shear stress on hepatic endothelium, with the result being nitric oxide generation and triggering of the hepatic regeneration cascade (Chapter 15).

2.1 MICROCIRCULATION

The microvascular unit of the liver is the hepatic acinus (Figure 2.3). The acinus represents a cluster of parenchymal cells approximately 2 mm in diameter. The parenchymal cells are grouped around terminal branches of the hepatic arteriole and portal venule [305,309]. The acini have been likened to clusters of berries suspended on a vascular stalk. This analogy is particularly appropriate because the vascular stalk enters the center of the acinus where the hepatic arterial blood and portal venous blood are well mixed within the sinusoidal periportal zone (Rappaport's zone 1). Flow in adjacent sinusoids is concurrent; all entrances to the acinus occur in the periportal region, whereas all exits occur at the periphery, thus producing strong gradients for oxygen and other substances that are added to or removed from the blood as it passes through the acinus. The central zone has the highest degree of oxygenation and the highest activity of respiratory enzymes [100,361]. Zone 3 lies on the outer limits of the acinus and is supplied by blood that has already passed the parenchymal cells of zones 1 and 2. Zone 3 is supplied by the least oxygenated blood and is rich in microsomal enzymes. This unique one-way flow arrangement precludes substances diffusing from the hepatic venous blood to the hepatic arterial resistance vessels. Therefore, even if the hepatic parenchymal cells release large quantities of vasoactive metabolites, such as adenosine [237], the parenchymal cells are not capable of regulating the hepatic artery according to their metabolic requirements.

Despite my personal preference for the acinus model, it should be emphasized that the classic lobule is more readily seen in the pig, where the lobule is bordered by connective tissue to form a distinct visual unit. This is not seen in humans or most other mammals (reviewed by Jungermann and Katz [157]).

2.2 HEPATIC MICROVASCULAR ZONES

All hepatocytes are not geared to perform identical functions. At any given time, metabolic events occurring in a single cell need not correspond with events in any other hepatocyte. When reports of gross analysis in the liver of a fed animal indicate that glycogen accounts for 5% and fat 4% of total cell volume, the impression is that every hepatocyte reflects these percentages. However, hepatocytes do not have the same percentage of stored fuels and the enzymes that control the rate of synthesis and utilization are not uniform among cells. The location and temporal differentiation of functions among the cells is called "metabolic zonation." Metabolic zonation can occur as a result of the oxygen gradient across the acinus or the gradient of hormones and nutrients that are extracted as the blood passes through the acinus, or due to chemicals added progressively to the blood before

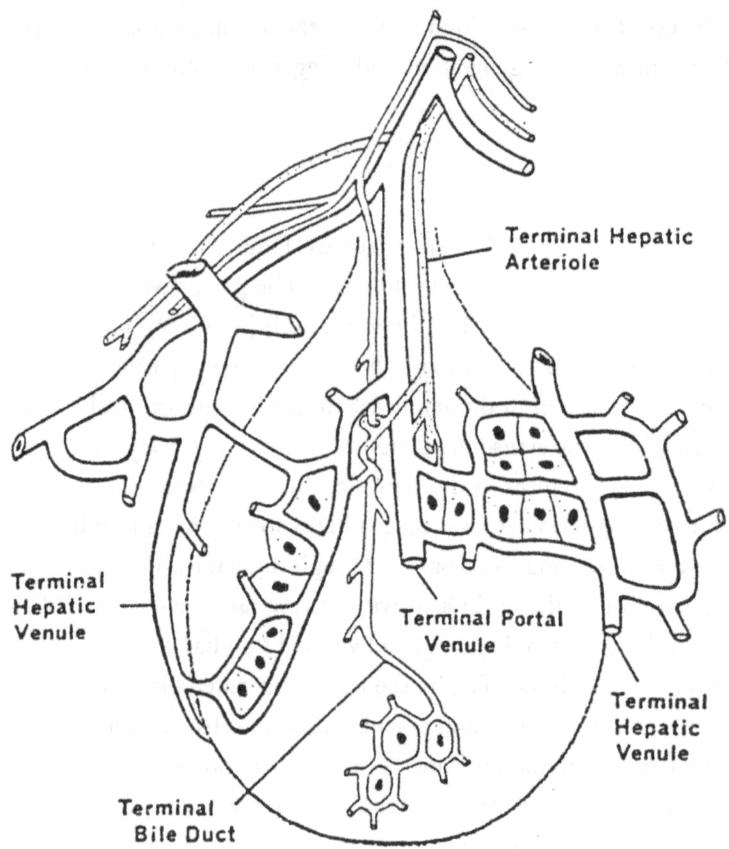

FIGURE 2.3: The acinus is the functional unit of the liver. There are approximately 100,000 acini per human liver; each is approximately 2 mm in diameter. Acini cluster like grapes at the end of vascular stalks comprising the terminal branches of portal veins, hepatic arteries, and bile ducts. Blood flows into the center (zone 1) of the acinus and flows outward to drain into terminal venules at the periphery (zone 3). Zone 1 is well oxygenated and is rich in nutrients, hormones, and toxins. Because flow in adjacent sinusoids is concurrent, zone 3 has the lowest oxygenation and short-circuiting of substances across the vascular system does not occur, nor do vasoactive substances in zone 3 diffuse back upstream to zone 1 where the hepatic arterial resistance vessels exist. There is no visible separation of sinusoids from zone 3 of adjacent acini. Based on the acinar concept of Rappaport (1973). Reproduced with permission from Lautt WW, Greenway CV. Conceptual review of the hepatic vascular bed. *Hepatology* 7: pp. 952–963, 1987. (This figure from publication Hepatology is reproduced with permission from publisher Wiley).

exiting the acinus. The partial pressure of oxygen in rat livers has been estimated at approximately 65 mmHg in the periportal zone (zone 1) with pO_2 in the perivenous zone (zone 3) of approximately 30–35 mmHg [360,379]. Oxidative energy metabolism is also predominant in zone 1. Enzymes of the tricarboxcylic acid cycle and respiratory chain are mainly located in hepatocytes in the periportal zone. Enzymes of glycolysis are relatively more active in the perivenous zone where oxygen tension is lower. The hepatocytes in the periportal zone are more devoted to gluconeogenesis than are those in the perivenous zone. The role of oxygen in determining these gradients is shown in studies on

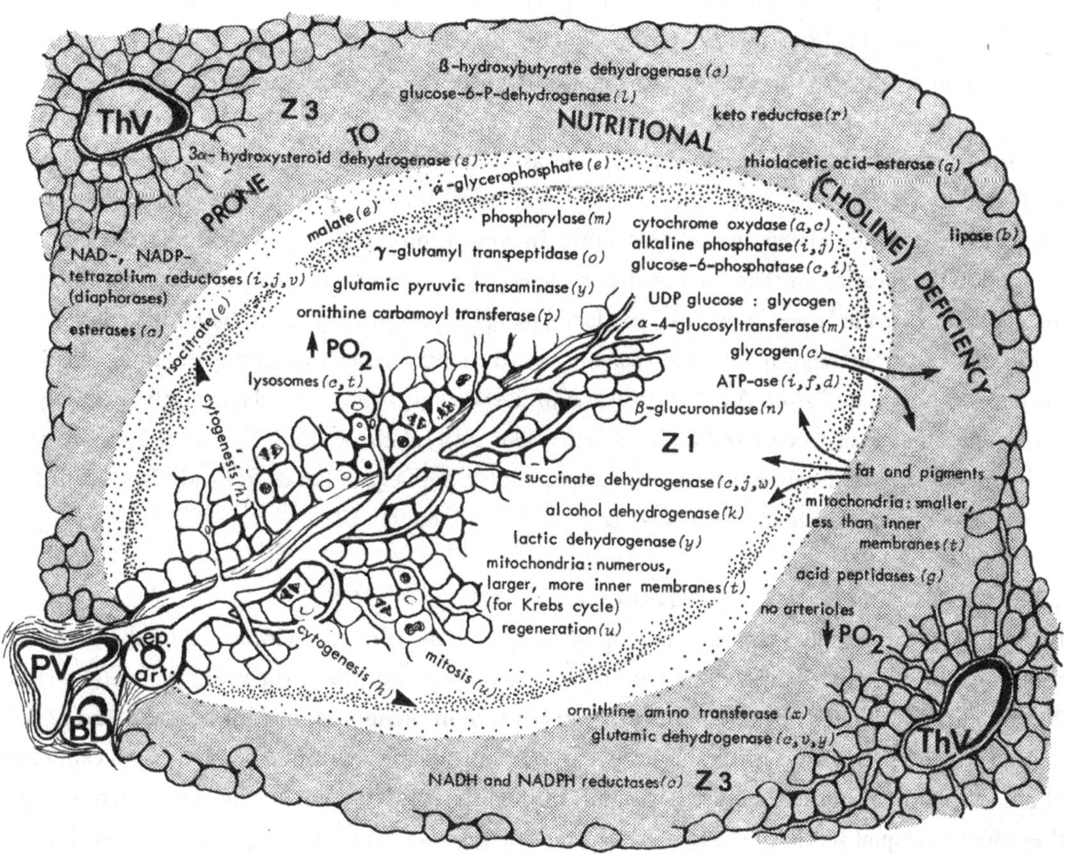

FIGURE 2.4: Metabolic areas in the acini. The central region (zone 1) is in the center of the acinus where oxygen content is the highest (white periportal area) and decreases progressively to zone 3 (shaded area). PV, portal vein; ThV, terminal hepatic venule; BD, bile ductule; hep. art., hepatic arteriole; Z1, zone 1 or periportal area; Z3, zone 3 or perivenous zone. Reproduced from Rappaport AM. In: *Diseases of the Liver*, Fourth Edition, edited by Schiff L. Philadelphia: Lippincott, Chapter 1, p. 10, 1975. (This figure from publication Diseases of the Liver is reproduced with permission of publisher Lippincott Williams & Wilkins).

isolated hepatocytes and on livers of rats that were perfused in a retrograde fashion, with the result that the metabolic zonation could be reversed in vivo [258].

Acute changes in hepatic circulation can lead to rapid changes in hepatocyte enzyme activity. The activity of γ-glutamyl transpeptidase is usually limited to the highly arterialized zone 1 in rat livers. Arterialization of the liver, by diverting portal blood, distributes the activity of the enzyme over all zones of the acinus [283]. How rapidly the changes occur is debatable but may suggest an additional potential role or effect of the hepatic arterial buffer response (Chapter 5). Figure 2.4 shows enzymatic activities that are predominant in each of the microcirculatory zones of the acinus.

The metabolic consequences of zonation have been reviewed and are beyond the purview of this monograph [100,128,158,309,361].

Cytotoxic injury is also subject to zonal differences [302]. Zonal necrosis is found most frequently in zone 3. The zonality appears to be related to the mechanism of injury, with zone 3 necrosis being induced by carbon tetrachloride, bromobenzene, and acetaminophen and has been attributed to the zonal concentration of the enzyme system responsible for the conversions of the agents to hepatotoxic metabolites. The necrosis in zone 1, produced by allyl formate, has been attributed to the location in that zone of the enzyme system that converts the compound to its toxic metabolite. Midzone (zone 2) necrosis produced by ngaione is attributed to the midzonal accumulation of its toxic metabolite. The necrosis due to idiosyncrasy-dependent hepatic injury in most instances is not zonal [387]. As fibrosis forms in livers subject to zone 3 damage, the histological pattern becomes easily recognizable as a starfish-like pattern surrounding the hepatic venule with each arm of the pattern representing the abutment of the perivenous zones from adjacent acini. The impact of zonation on hepatic toxicology is discussed in Chapter 12.

2.3 INTRAHEPATIC FLOW DISTRIBUTION

Blood flow within the liver appears to be quite uniformly distributed, as indicated by the even distribution of microspheres injected into either the hepatic artery or the portal vein [119]. The surface 2 mm of the liver directly beneath Glisson's capsule is more richly supplied by arterial blood [232].

Intraportal and intra-arterial infusions of norepinephrine result in R_{max} and ED_{50} estimates of blood volume responses that are not significantly different. Norepinephrine administered by either route has equal access to the hepatic capacitance vessels [231]. The use of radioactively tagged microsphere distribution throughout core samples taken from top to bottom of liver lobes administered at several different time points demonstrated a dynamically changing vascular perfusion. The heterogeneity of flow distribution decreased in response to norepinephrine in contrast to the increase in heterogeneity seen in response to norepinephrine in isolated liver preparations [232].

Substances reaching the liver via the hepatic artery or portal vein are equally well extracted, suggesting that both vascular inlets perfuse the hepatic parenchymal cells equally [221]. Bile salts produce equal stimulation of bile flow, despite producing quite different direct vascular effects, when infused via the two vascular channels [209]. As an approximation, the hepatic arterial and portal venous flows meld in equal proportions throughout the liver. Elevation in venous pressure, reduction of portal venous flow, and stimulation of hepatic nerves or norepinephrine infusion do not result in flow redistribution within the liver [59,120,128]. We found no gravity effect on the ratio of arterial to portal blood flow within the cat liver [232].

In the normal liver, it seems likely that one function of the hepatic arterial buffer response (Chapter 5) is to maintain homogeneity of liver perfusion. If local portal venous stasis occurs, the hepatic artery supplying that acinus should dilate to increase the higher pressure arterial input, thereby flushing the sinusoids and restoring sinusoidal blood flow.

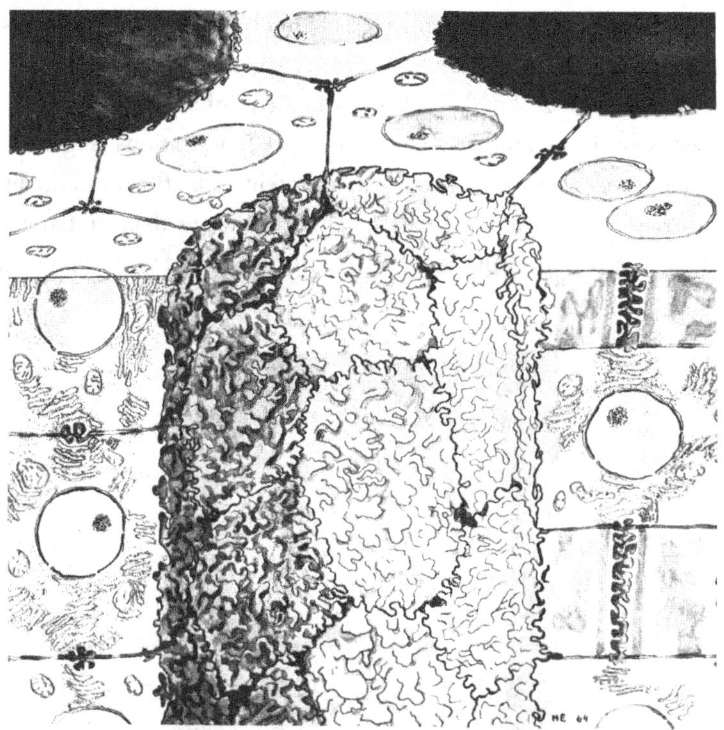

FIGURE 2.5: Hepatocytes are connected as a syncytium of cell plates of one cell thick. The syncytium is tunneled through with lacunae that are defined by the space between hepatocyte plates and contain the space of Disse (equivalent to extracellular space in other tissues), the fenestrated endothelial cells (equivalent to capillary endothelium in other tissues), and the plasma volume. Basement membrane structures, Kupffer cells, and Stellate cells lie within the lacunae. The microvilli of the parenchymal cells extend into the space of Disse. Bile canaliculi form between contacting hepatocytes and gap junctions facilitate cell-to-cell communication. Reproduced from Elias H, Sherrick JC. *Morphology of the Liver*. Academic Press, New York, New York. Figure I-15, p. 31, 1969. (Figure I-15 from publication Morphology of the Liver is reproduced with permission of publisher Academic Press).

Blood and solutes enter the sinusoidal microcirculation via the terminal portal venules or terminal hepatic arterioles in the central or zone 1 of the liver acinus [307]. Sinusoids distribute the blood sequentially through acinar zones 1, 2, and 3, passing approximately 16–20 hepatocytes before terminating in hepatic venules at the acinar periphery. Red blood cells remain restricted within the sinusoidal space defined by the endothelial cells, which have large fenestrations and permit molecules as large as albumin to pass through the fenestrations and enter the small space of Disse before making contact with the microvilli of the hepatocytes. The volume of the sinusoids in zone 1 is less than zone 3 but the surface area per unit volume is higher in zone 1, thereby facilitating uptake of compounds from the space of Disse [268].

The hepatic sinusoids are arranged in a syncytium of interconnected spaces referred to as hepatic lacunae. The lacunae are the interconnected channels between the sinusoids with all but the parenchymal cells removed (Figure 2.5). The lacunar space consists of the plasma compartment and the space of Disse, separated and defined by a sinusoidal endothelial cell layer. Within the lacunar space lays a continuous layer of endothelial cells with large fenestrations (Figure 2.6). When Starling elevated the hepatic venous pressure and noted a protein-rich effluent in the hepatic lymph, he deduced the characteristics of the sieve as being much more porous than similar endothelial cells in tissues such as muscle. Chronic liver disease is frequently associated with capillarization of the he-

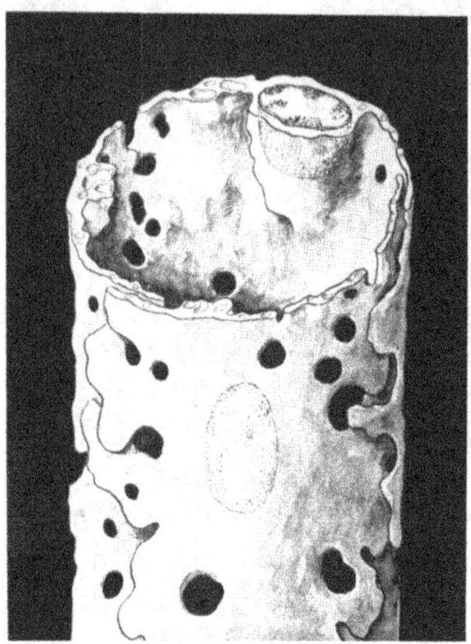

FIGURE 2.6: The sinusoidal endothelial cells form a tube within the hepatic lacunae. The large fenestrations permit passage of large molecules between the space of Disse and the plasma. The pores are much smaller in muscle and absent in brain. Reproduced from by Elias H, Sherrick JC. *Morphology of the Liver*. Academic Press, New York, New York. Figure I-16, p. 31, 1969. (Figure I-16 from publication Morphology of the Liver is reproduced with permission of publisher Academic Press).

patic endothelial cells with reduced fenestrations and, subsequently, reduced freedom of movement between hepatocytes and plasma. The space of Disse is a small space, equivalent to the interstitial fluid of other vascular beds.

2.4 KUPFFER CELLS

Kupffer cells are liver-specific macrophages that reside in the sinusoidal lumen where they are exposed to the systemic circulation via the hepatic artery and to the splanchnic circulation via the portal vein. They constitute approximately 80% of the total population of macrophages in the body [291]. Because of their phagocytic capacity, they participate in the host defense system to clear circulating endotoxin from the blood [22,376]. Kupffer cells are also capable of secreting mediators involved in the host responses to inflammation, such as cytokines, endothelins, prostanoids, and nitric oxide [53,323].

Kupffer cells activate inducible nitric oxide synthase, producing large amounts of nitric oxide [348]. This enzyme is not present in resting cells and, after stimulation, requires a period of mRNA expression and new protein synthesis to detect enzyme activity [23,348]. Different stimuli such as cytokines, interferon-γ, lipopolysaccharide, tumor necrosis factor, and interleukin-1 are able to induce the synthesis and to release nitric oxide from Kupffer cells [66,177].

Kupffer cells not only play an important role in the body's defense system but may also contribute to liver damage [376]. Extensive nitric oxide production by Kupffer cells causes cytotoxicity. Nitric oxide damaging effects might be due to a cooperative action with superoxide, yielding the peroxynitrite anion (ONNO–). Peroxynitrite, known to oxidize sulfhydryls and to generate products indicative of hydroxyl radical reaction with deoxyribose and dimethyl sulfoxide, induces lipid peroxidation [18,304]. Nitric oxide cytotoxic effects are dependent on the concentration of superoxide radical [328]. In chronic alcoholic rats, production of both superoxide radical and nitric oxide in isolated Kupffer cells seems to be at the same rate [13,370]. Similarly, the superoxide radical from mitrochrondria and microsomal enzymes increases as nitric oxide concentrations increase in hepatocytes from alcoholic LPS-treated rats [28,173,370].

Nitric oxide by itself might have a protective effect by regulating the production of specific inflammatory mediators by Kupffer cells during sepsis [350]. In the septic liver, nitric oxide has a profound inhibitory effect on the production of prostaglandin E_2 (PGE_2), thromboxane B_2, and interleukin-6 [350]. In this case, nitric oxide might be an autoregulator of inflammatory reactions. Nitric oxide and superoxide radicals are used as weapons by the Kupffer cells to kill engulfed microorganisms. Macrophage use of these weapons is not unique to the liver. Macrophages, including Kupffer cells, attack, engulf, and kill invading organisms.

Kupffer cells can have an influence on microvascular flow by protruding their pseudopods toward the vascular space in response to endotoxin administration in vivo [260,289]. Moreover, the nitric oxide produced by Kupffer cells may affect microvascular flow because of the strong vasorelaxant properties acting on other cells (Chapter 9). Indirectly, nitric oxide, by suppressing Kupffer cell eicosanoid synthesis, may also have an impact on vascular tone. However, keep in mind the anatomy of the acinus. Vasoactive substances released from Kupffer cells cannot diffuse upstream to act on the resistance vessels of the terminal branches of the hepatic artery and portal vein.

Kupffer cells straddle inside the sinusoidal space like a spider in ambush for particulate substances passing through the hepatic circulation. Kupffer cells account for approximately 15% of the liver cell population. They are attached by their pseudopodia to the endothelial cells and some of these processes penetrate through the fenestrae and reach out to the parenchymal liver cells [377]. Although Kupffer cells are generally considered to be fixed macrophages, in vivo microscopy shows that they can independently move at a speed of 4.6 μm/min [253]. Kupffer cells are more abundant in zone 1 of the acinus (43% of total) and those cells are larger and more active in phagocytosis. Zone 2 contains 32% and zone 3, 25% of total Kupffer cells [174] with the smaller perivenous Kupffer cells appearing to be more active in cytokine production and to have a higher cytotoxic capacity [174].

Twenty-five percent of the cardiac output passes through the liver, thereby allowing an efficient filtration role for the macrophage activities of the Kupffer cell. No direct vascular role has been suggested for the Kupffer cell (see Chapter 9 for vascular role of nitric oxide and Chapter 15 for role of nitric oxide in the regulation of hepatocyte proliferation).

2.5 STELLATE CELLS

Stellate cells, also known as fat-storing cells, Ito cells, or hepatic perisinusoidal lipocytes, are mesenchymal cells that reside in the perisinusoidal space of Disse and can undergo reversible contraction [160,320]. Intrahepatic resistance and portal pressure are affected by stellate cells [160,162]. They express smooth muscle-specific intermediate filament desmin, which allows these cells to contract [319]. Stellate cells generally form around the exterior of the endothelial cells like fingers that are capable of compressing the sinusoidal diameter by squeezing the endothelial cells [383].

Studies on isolated stellate cells showed that vasoconstrictors, such as angiotensin II, thrombin, and endothelin-1, increase intracellular free calcium, which is coupled with cell contraction [301]. Kawada et al. [161] quantified constriction and relaxation of stellate cells by culturing the cells on a flexible silicone rubber membrane to measure tension. Constriction of the stellate cells was elicited by a thromboxane A_2 analogue, prostaglandin $F_{2\alpha}$, and endothelin-1, whereas prostacyclin (prostaglandin I_2) analogues and PGE_2 induced cell relaxation [160,161]. Sinusoidal constriction induced by endothelin-1 may be mediated by stellate cells; this effect was inhibited by nitric oxide donors [383]. It is suggested that a balance between dilators and constrictors becomes disrupted in conditions of reperfusion injury or endotoxin shock, and that the balance favors stellate cell constriction, thus leading to sinusoidal regional blood flow heterogeneity [54].

Sinusoidal blood pressure and vascular resistance are so low that a pressure gradient across the liver from the portal venous inflow to the hepatic venous outflow is only approximately 5 mmHg. The low pressure gradient is remarkable considering that 30% of the inflow to the liver sinusoids is provided by the hepatic artery under arterial pressure. With such a low sinusoidal perfusion pressure, small imbalances in microscopic sinusoidal flow could lead to stagnation at many sites. The hepatic arterial buffer response works at the acinar level. The regulation of flow within the acinus may also be actively controlled by regional regulators at a single-cell level. The stellate cells likely play such a role, although this does not seem to have been shown. It would seem that the stellate cell would have to detect zero flow and then dilate to pull outward on the endothelial cells and enlarge the sinusoidal space.

2.6 SINUSOIDAL ENDOTHELIAL CELLS

Injury of sinusoidal microvasculature is one of the first events in the sequel of developing hepatic failure during severe sepsis, endotoxemia, or reperfusion injury [136]. The inducible nitric oxide synthase is activated in the liver in inflammatory conditions [320,349]. The affected sinusoidal endothelial cells might lead to alterations in nitric oxide levels. Increased release of nitric oxide by Kupffer cells could be anticipated to have an impact on sinusoidal blood flow by directly relaxing stellate cells or smooth muscle [321]. However, nitric oxide released by hepatic sinusoidal cells is unlikely to influence the arterial or portal resistance site upstream. Nitric oxide produced by the endothelium of the inlet vessels does, however, play a significant vascular role (Chapters 5 and 9).

A role of sinusoidal endothelial nitric oxide may be minor for the vasculature, but a key role for regulating hepatocyte proliferation is described in Chapter 15. A role for vascular endothelium in both the hepatic artery and the portal vein is described in Chapter 5.

Nitric oxide also plays a major regulatory role in glucose homeostasis. Dysfunction of hepatic parasympathetic nerve-induced stimulation of nitric oxide synthase results in insulin resistance in skeletal muscle and is suggested as the initiating metabolic defect in the prediabetic state that leads progressively to obesity, syndrome X, and type 2 diabetes. This science is discussed elsewhere (for a review, see reference [202]).

2.7 THE SPACE OF MALL

The space of Mall, variously known as the portal duct or portal triad, represents the service conduit that supplies the center of each acinus with portal and arterial blood, collects bile and lymph, and through which passes a rich array of sensory and efferent nerves. The terminal branches of the hepatic artery are in intimate contact with the terminal branches of the portal vein. Direct arterial branches have been shown to enter the terminal portal venules before entry into the sinusoids. An additional vascular circuit exists within the space of Mall, the peribiliary plexus (Figure 2.7).

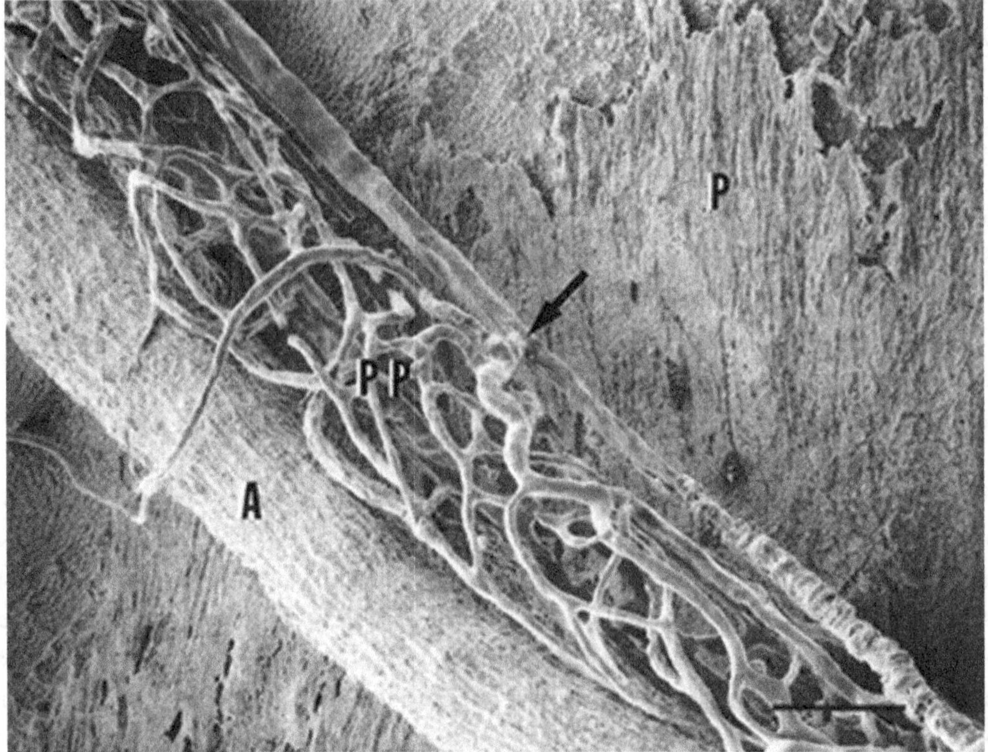

FIGURE 2.7: A cast of the portal vein (P), hepatic artery (A), and peribiliary arterial plexus (PP) of a rat, showing a connection between a small artery and the plexus (arrow). The peribiliary plexus forms a dense sheath around the bile duct, suggesting function interaction. Bar = 100 μm. Reproduced from Grisham JW, Nopanitaya W. Scanning electron microscopy of casts of hepatic microvessels: review of methods and results. In: *Hepatic Circulation in Health and Disease*. New York: Raven Press, Figure 4, p. 98, 1981. (Figure 4 from publication Hepatic Circulation in Health and Disease is reproduced with permission of publisher Raven Press).

The hepatic artery sends off small branches that form a dense mesh-like plexus surrounding the bile duct. A significant proportion of the hepatic arterial flow passes through the peribiliary plexus before it drains into the portal vein at or near where the portal vein and hepatic artery merge at the origin of sinusoids. The function of the peribiliary plexus is not known, although substances such as adenosine, when back-perfused through the bile duct, reach the hepatic arterial resistance vessels and cause dilation.

The fluid in the space of Mall is also proposed to serve as a medium for signals between hepatic blood vessels (Chapter 5) and sensory nerves (Chapter 13).

• • • •

CHAPTER 3

Fluid Exchange

Child [52] reviewed the history of the debates regarding the movement of fluid from the sinusoidal space through the space of Disse, the space of Mall, and into hepatic lymphatics, with free access through the Glisson's capsule on the surface of the liver. The morphological detail of the passage of fluid through these sites is no clearer than it was in 1954. Regardless of the precise morphology and arguments as to whether these compartments are actually connected, the observations are that lymph of high-protein content can be forced from the plasma compartment of the sinusoid to exit the liver through either hepatic lymphatic outflow or weeping from the surface of the liver. The free passage of high-protein fluid is readily demonstrated through these morphological structures. Starling, in 1896, appears to have been the first to specifically elevate hepatic venous pressure and observe a high-protein, high-volume lymphatic outflow from the liver. If the lymph outflow is blocked, the filtered fluids exude across the surface of the liver. The droplets on the surface of the liver have a protein content similar to that of plasma. This process can be visualized by encasing the liver in a plethysmograph filled with mineral oil. Every drop of exudate becomes visible as a growing tear. The fluid exudation that occurs across the surface of the liver under the influence of increased hepatic venous pressure has a protein content 80% of that of the plasma.

The volume of lymph secreted per 24 h in a dog amounts to 47% of the estimated total plasma volume. This fluid volume contains 35% of the total circulating plasma proteins. In experimental cirrhosis, the flow of hepatic lymph has been estimated to increase as much as 258% and, over a 24-h period, contains as much as 207% of the circulating plasma proteins (Child, 1954). Total lymph flow from the liver apparently has not been measured. Brauer [30] estimated it at 0.04–0.06 ml/min per 100 g of liver, whereas Laine et al. [180] reported 0.06 ml/min from one prenodal lymphatic in dogs.

The role of hepatic lymph is still not clear. The lymphatic drainage in tissues like skeletal muscle serves to maintain the interstitial fluid volume and hydrostatic pressure at a low level and to salvage proteins and other large molecules that have escaped the vasculature. The liver requires neither of these functions. There is virtually no hydrostatic pressure gradient or colloid osmotic pressure gradient being exerted across the fenestrated endothelial cells. The space of Disse, the hepatic version of interstitial space, is freely connected to the plasma space. There is no interstitial space that requires lymphatic pickup. So why does the liver have such a huge lymph flow? Perhaps there is a value to secretion of newly formed hepatic proteins into a lymphatic system to be delivered to the body through the thoracic lymph flow. However, there would appear to be little advantage compared to simply secreting the protein directly into the space of Disse where it will be quickly washed away into the sinusoidal space and the systemic circulation. It is more probable that the fenestrations restrict mainly lipoproteins, including the large very low-density lipoproteins that are secreted by the liver. In diseased livers, where fenestrations become reduced in size and number, the liver could still secrete triglycerides for transport to adipose tissue by very low-density lipoprotein passage through lymph, whereas uptake of cholesterol from low-density lipoprotein would be restricted (Chapter 12).

A linear relationship between hepatic venous pressure and hepatic lymph flow was shown in cats by Neil Granger and collaborators [101] and in dogs by his brother Harris and collaborators [180]. By calculating the ratio of proteins of different molecular weights in the plasma and the hepatic lymph, it was shown that the appearance in lymph was highly dependent on the molecular weight. However, when hepatic venous pressure was elevated, free passage of large and small molecules was observed [101]. Molecules the size of albumin appear at 95% of the concentration of that in plasma in rabbits. In all species where hepatic lymph has been collected, the amount of protein is 76–95% that seen in plasma [12]. Whatever mild restriction on fluid motion may exist, elevation of sinusoidal pressure forces all channels open so that even large molecules are filtered through the lymphatics.

A major source by which fluid from the plasma rapidly exits the liver, in conditions of elevated sinusoidal blood pressure, is directly through the surface of the liver through a lymphatic network that has unclear connections to the lymphatics deeper within the tissue. By excluding hepatic lymphatic efflux by ligating the lymph vessels, all fluid filtered from the sinusoidal space can be quantified by measurement of hepatic surface exudates or, much more conveniently, by measuring changes in filtration rates through the use of an in vivo plethysmograph [103].

Figure 3.1 shows the blood volume and fluid filtration response to an acute elevation in hepatic venous pressure of 9.4 mmHg for 60 min. Immediately upon elevation of hepatic venous pressure, the liver volume underwent a rapid expansion, increasing by 20 ml per 100 g of liver within 1 min. This rapid increase in volume was entirely due to an expansion of the hepatic blood volume. As the hepatic venous pressure was maintained, volume within the plethysmograph continued to increase and by 20 min was increasing at a linear rate that continued for the full duration of the increased pressure. Filtration began immediately and continued at a constant rate as long as the venous pressure was elevated (tested up to 6 h). Reducing the venous pressure led to a rapid expulsion of blood volume to previous precongestion levels.

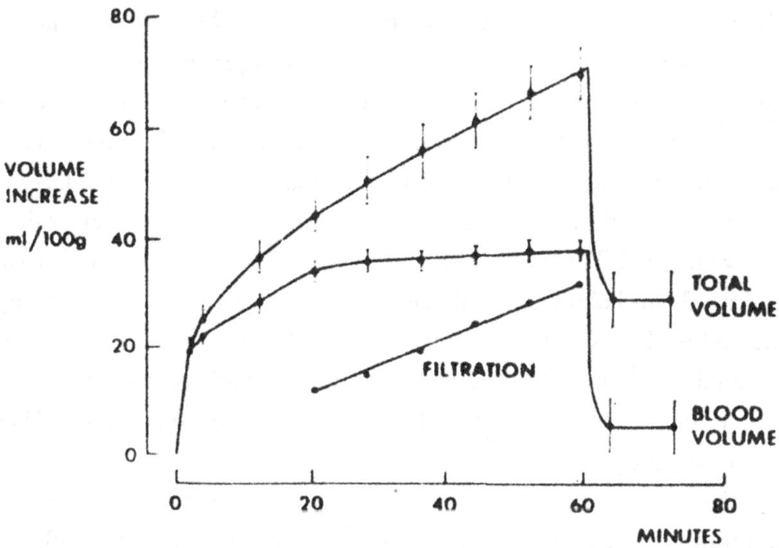

FIGURE 3.1: Changes in total hepatic volume and blood volume (mean ± SE) in cats when hepatic venous pressure was increased to 9.4 mmHg for 60 min. Blood volume changes were detected using Cr^{51}-tagged red blood cells and radioactive emissions from the liver in the plethysmograph. Transsinusoidal fluid filtration was calculated by subtracting mean values of two curves. Reproduced from Greenway CV, Lautt WW. Effects of hepatic venous pressure on transsinusoidal fluid transfer in the liver of the anesthetized cat. Circ Res 26: pp. 697–703, 1970. (This figure from publication Circ Res is reproduced with permission from publisher).

3.1 FLOW-LIMITED DISTRIBUTION OF BLOOD-BORNE SUBSTANCES

The "multiple indicator dilution technique" consists of determining the hepatic venous efflux after rapid intraportal injection of a reference substance, whose behavior is known, and one or more other substances, whose behavior is to be characterized. One of the pioneers who used the technique for studies in the liver was Carl Goresky. Carl contributed a chapter in the 1981 symposium that provides a theoretical basis for the method and its interpretation [97]. For this method, red blood cells were usually used as the reference curve. A labeled red cell travels faster than a bolus of labeled albumin. The difference in rate of efflux from the liver represents different accessibility to the hepatic spaces. Access to intracellular space is restricted according to molecular weight. The restriction by molecular size demonstrated by the indicator dilution outflow curves from the liver is compatible with the completely different approach of demonstrating physical size limitation to hepatic lymph outflow [12].

On the basis of a series of studies with multiple-indicator dilution techniques, Goresky [98] confirmed that no continuous anatomic barrier exists between plasma and the space of Disse. The endothelial lining serves to contain red cells, but virtually immediate lateral diffusion equilibrium is expected for most molecules within the space of Disse. As blood cells squeeze through the sinusoids, they massage the endothelial cells and further mix plasma and Disse fluid.

Sinusoids in zone 1 of the acinus form richly anastomotic channels, whereas sinusoids in zone 3 are more radially arranged. There are fenestrations throughout the length of the sinusoids composed of single large 3-μm holes and small clusters of 1-μm holes surrounded by microfilaments [128]. There are no lymphatics associated with the sinusoids. Lymph vessels appear to arise in the space of Mall around the portal tract, forming extensive lymphatic plexuses that anastomose with other lymphatic plexuses on the surface of the liver underneath Glisson's capsule. These anastomoses thus provide a lymphatic pathway from deep in the liver to the surface. Other lymphatic plexuses are found around the larger hepatic veins [30,57]. When formation of lymph exceeds the capacity of the major lymph trunks, exudation of lymph on the surface of the liver occurs [30,31,110,141].

Morphological studies suggest that the sinusoidal lumen is 10.6% of liver volume, the space of Disse is 5%, and the bile canaliculi comprise 0.4% of volume [24]. The picture that emerges from this anatomy is compatible with the physiological data on extracellular spaces in the liver. Sodium (8.9%), sucrose (8.8%), and inulin (7.7%) spaces were almost equal. Albumin extravascular space was 5.7% of liver weight and 64% of the Na^+ space (98). The albumin space was 59% of the interstitial space, and this increased to 66% when hepatic venous pressure was raised. The lymph-to-plasma concentration ratio varied only slightly with molecular size and was 1.0 for lactoglobulin, 0.88 for albumin, and 0.69 for γ-globulin. Protein content of lymph was 80–95% that of plasma, and hepatic lymph proteins originated from blood, not from new synthesis [30,71,101,122,180].

These data indicate that 60% of the interstitial space is accessible to albumin, and this presumably includes the spaces of Disse, the spaces of Mall, and the lymphatics. Lymph appears to be formed by filtration of high-protein fluid from the spaces of Disse across the limiting plate into the lymphatics in the spaces of Mall, and across the limiting plate into the tissue spaces and lymphatics around the hepatic veins. The limiting plate appears to represent a very minor barrier to passage of proteins, depending on their molecular weights [71,101,180].

3.2 ASCITES FORMATION

Ascites is a collection of fluid in the peritoneal cavity that represents an imbalance between the rate of fluid filtration into the peritoneal space and the rate of reabsorption from that space. The liver does not become edematous. Fluids filtered out of the sinusoidal space rapidly exit with apparently minor restriction. There is no evidence that fluid reabsorption from the peritoneal space through the liver occurs, nor do the characteristics of the vasculature suggest this would be possible.

Zink and Greenway [386] studied the reabsorption of fluid from the peritoneal cavity by encasing the abdomen in a rigid plaster cast to form an abdominal plethysmograph. The rate of fluid absorption from the peritoneal cavity was directly proportional to the intraperitoneal pressure regardless of whether the intraperitoneal fluid was free from protein or contained a protein concentration equivalent to that of plasma. The relationship between absorption of fluid from the peritoneal space and intraperitoneal pressure was approximately linear with a rate of 0.01 ml min^{-1} $mmHg^{-1}$ per kilogram of body weight or 0.04 ml min^{-1} $mmHg^{-1}$ per 100 g of liver. This contrasts with the filtration rate out of the liver of 0.08 ml min^{-1} $mmHg^{-1}$ per 100 g of liver [103]. Thus, removal from the peritoneal cavity per unit pressure appears to be substantially slower than formation, and this may explain the occurrence of ascites in some pathological situations. Further studies with ^{131}I-labeled albumin showed that protein was absorbed from the peritoneal cavity in equal proportion to the absorption of fluid, and the fractional rates of protein absorption were never significantly different from the fractional rates of fluid absorption. Both fractional rates were independent of the protein concentration in the peritoneal cavity, and this indicated that the removal process involved lymphatic absorption rather than transcapillary absorption [261].

It is very difficult to reverse the hydrostatic pressure gradient across the small hepatic vessels because raising intraperitoneal pressure compresses the large veins. Thus, sinusoidal pressure increases by the same amount as extravascular pressure [385]. It appears that protection of the liver against edema is achieved by lymphatic drainage and transudation across the capsule rather than by mechanisms that limit filtration or that allow reabsorption of the fluid into the hepatic vessels [110,180].

3.3 EFFECTS OF DRUGS ON FLUID EXCHANGE

When hepatic venous pressure was mildly elevated to produce a steady-state filtration across the liver, infusions of epinephrine, isoproterenol, and histamine had no effect on the steady-state filtration, and it was concluded that these drugs did not modify either surface area or permeability within the liver [111]. There are clear species differences in this regard as histamine is capable of stimulating vasoconstriction in the hepatic veins of dogs, thereby increasing intrahepatic pressure and fluid filtration [19] (also see Figure 6.5 in Chapter 6).

3.4 EFFECTS OF HEPATIC NERVE STIMULATION

Some of the earlier responses of fluid exchange in response to vasoactive stimuli were misinterpreted based on the assumption that central venous pressure was an estimate for intrahepatic sinusoidal pressure. Vasoconstriction resulting in elevated portal venous pressure was thought to primarily occur at the presinusoidal portal vascular resistance vessels. Vasoconstrictors have since been shown to have the primary effect in small hepatic venules so that portal venous pressure is more reflective of intrahepatic sinusoidal pressure (Chapter 6).

Regardless of the proportion of the vasoconstriction that occurs at presinusoidal or post-sinusoidal sites, stimulation of the hepatic sympathetic nerve branch will result in some degree of increase in sinusoidal pressure. Therefore, one should expect to see an increase in fluid filtration from the liver. This is not seen. In fact, if a background filtration is established by elevating hepatic venous outflow pressure to produce a portal pressure of 8.7 mmHg, hepatic nerve stimulation at 2, 4, and 8 Hz results in a frequency-dependent decrease in filtration rate. If hepatic venous pressure is then elevated to levels of 12 and 16 mmHg, the ability of nerve stimulation to decrease the large filtration rate is overwhelmed but the trends still remain, indicating that the nerves result in reduced fluid filtration even in the face of elevated sinusoidal pressure [105]. These studies suggest that the hepatic sympathetic nerve stimulation reduced fenestration size and impaired fluid filtration out of the plasma compartment.

Even if the entire vasoconstrictor response to the nerves was at the presinsuoidal site, sinusoidal pressure would not decrease to reduce fluid filtration. Hepatic blood flow is neither significantly redistributed nor heterogeneous when hepatic nerves are stimulated (Chapter 11). Therefore, heterogeneity of flow cannot account for the ability of the hepatic nerves to decrease fluid filtration. This observation requires further evaluation as it may have significant consequences for liver metabolism in situations of activated sympathetic nerves. Furthermore, if the endothelial fenestrations can be shown to be regulated by constrictor influences, then dilator stimuli should be identified for evaluation of utility in diseased livers where sinusoidal fenestrations become decreased in number and size. Access of lipoproteins transporting triglycerides from the liver (very low-density lipoprotein) and cholesterol to the liver (low-density lipoprotein) may be manipulated by regulating fenestrae size.

3.5 BLOOD FLOW AND HEPATIC CLEARANCE OF DRUGS AND HORMONES

As a general rule (with many important exceptions), clearances of substances through any organ have certain kinetic limitations. Generally, if a substance has a very high extraction ratio passing through the liver, a reduction in hepatic blood flow will lead to a similar reduction in hepatic clearance of that substance. That can have major homeostatic consequences. If substances have a very low hepatic clearance rate, it is assumed that the rate-limiting step is not the delivery of the compound to the extracting hepatocytes but rather a rate-limiting control within the processing cells. This simplistic pharmacokinetic model cannot be relied upon. For example, we showed that hepatic extraction of lidocaine in the cat was only 28%. At that moderate level of hepatic extraction, classic pharmacokinetic theory at that time would have predicted a very significant impact on hepatic extraction as a result of a reduction in hepatic blood flow. However, as hepatic blood flow decreased,

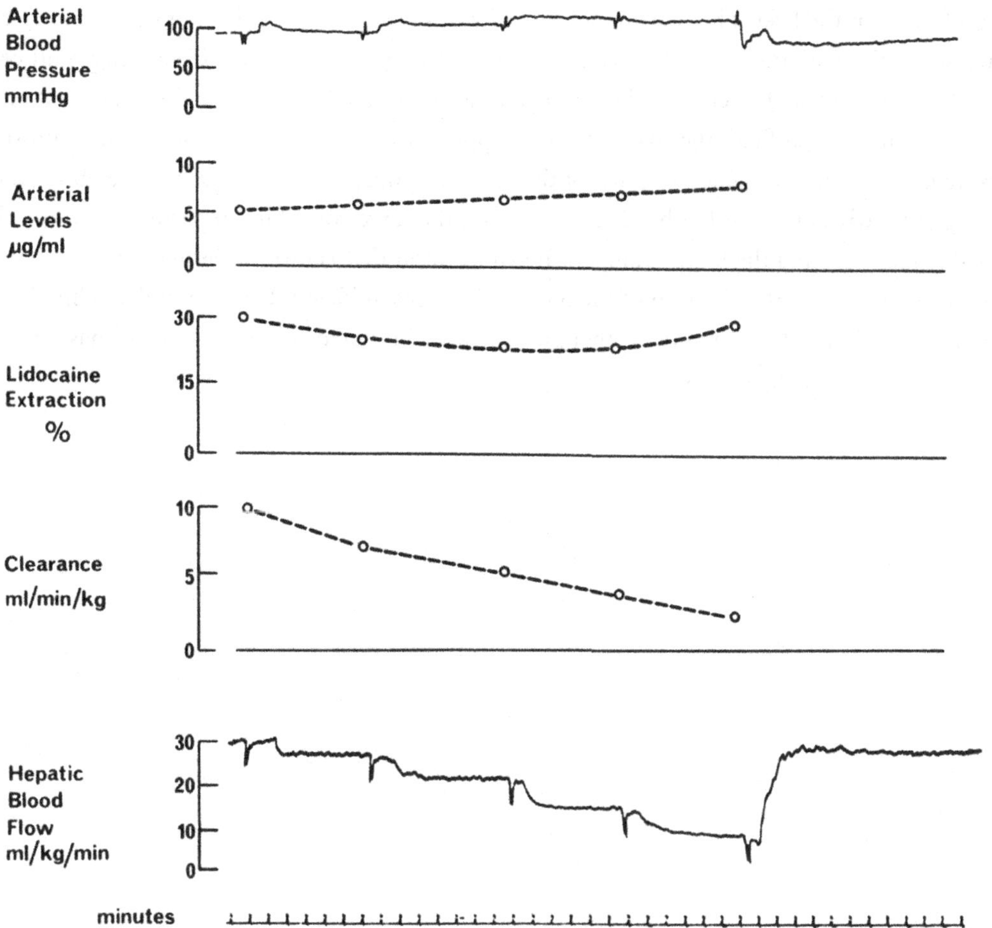

FIGURE 3.2: The effect of stepwise reduction in hepatic blood flow on lidocaine clearance rate. Control extraction ratio of 30% was not significantly altered by reduced flow. Lidocaine clearance is linearly related to blood flow in the cat. Reproduced with permission from Lautt WW, Skelton FS. The effect of SKF-525A and of altered hepatic blood flow on lidocaine clearance in the cat. *Can J Physiol Pharmacol* 55(1): pp. 7–12, 1977. © 2008 NRC Canada or its licensors. (This figure from publication Can J Physiol Pharmacol is reproduced with permission from publisher NRC Canada).

the hepatic extraction ratio for lidocaine did not change significantly and lidocaine clearance decreased in parallel with blood flow (Figure 3.2).

In summary, the hepatic lymphatic system is poorly understood in terms of function, regulation, and role. The pathway by which fluid, filtered from the plasma space, enters both the lymphatics that drain into the thoracic duct and into lymphatics, through which fluid exudates across Glisson's capsule, is unclear. The role of the lymphatic system in other tissues does not appear to be of significance in the liver. The Starling forces of hydrostatic pressure and colloid osmotic pressure acting across the endothelial cells to regulate fluid exchange between the extracellular fluid compartments of plasma and interstitial fluid seems irrelevant to the liver because the equivalent of the extracellular interstitial fluid, the space of Disse, appears to have a protein content and hydrostatic pressure level indistinguishable from that of the plasma compartment. Although these observations may suggest a trivial role for the hepatic lymphatics, the huge contribution of hepatic lymph flow to total lymph flow and the large volume of hepatic lymph that is formed per day suggest that our knowledge rather than the function of the lymphatics is insignificant. I cannot update Child's conclusion in 1954 that "At the moment, then, it can safely be stated that the last word has not been written on the hepatic lymphatics."

• • • •

CHAPTER 4
Capacitance

The integrity of the venous capacitance system is crucial to the determination of the overall effect of vasoactive compounds on cardiac output and systemic blood pressure. If peripheral resistance is decreased and the capacitance vessels are simultaneously constricted (elevated preload), the cardiac output will be markedly elevated, but arterial pressure may remain unaltered. If arterial and venous dilation occur simultaneously, cardiac output may remain unaltered, but arterial pressure will decline dramatically. Simultaneous contraction of arterial resistance vessels and the venous capacitance vessels will lead to large elevations in systemic blood pressure with minimal effects on cardiac output, whereas contractions of arterial resistance vessels with inactive capacitance vessels will result in reduced cardiac output. Approximately 25% of an acute increase or decrease in plasma volume is rapidly accommodated by the liver.

Visible acute changes in liver size can be seen without the intervention of technology. Liver volume is easily demonstrated to increase dramatically in response to elevated central venous pressure or venous outflow block. Increased liver volume that was engorged with blood was identified in congestive heart failure. The expulsion of a large volume of blood from the liver in emergency situations was also an early observation. In 1930, Griffith and Emery observed a decrease in liver volume when the hepatic sympathetic nerves were stimulated, although no quantitative data were reported. The adaptation of the classic plethysmograph to fit the intact functioning liver provided the opportunity for precise quantification of both blood volume and fluid filtration responses in large animals (cats and dogs).

I was a junior graduate student carrying out experiments in partnership with Ron Stark who was doing a postdoctoral year with Clive Greenway. Greenway conceived the idea for an in vivo plethysmograph from his encounters with Stefan Mellander and the Swedish research community (see Chapter 1 for historical perspective and Chapter 3 for fluid exchange).

Greenway [103] presented a detailed description of the construction and operation of the Plexiglas plethysmograph. The same design was modified for use in dogs, but attempts to miniaturize for rats were not successful (Greenway, personal communication).

Briefly, the liver, with the exception of the right posterior lobe, was inserted onto the base of a Plexiglas plethysmograph (Figure 4.1) after the ligaments connecting the central and left lobes

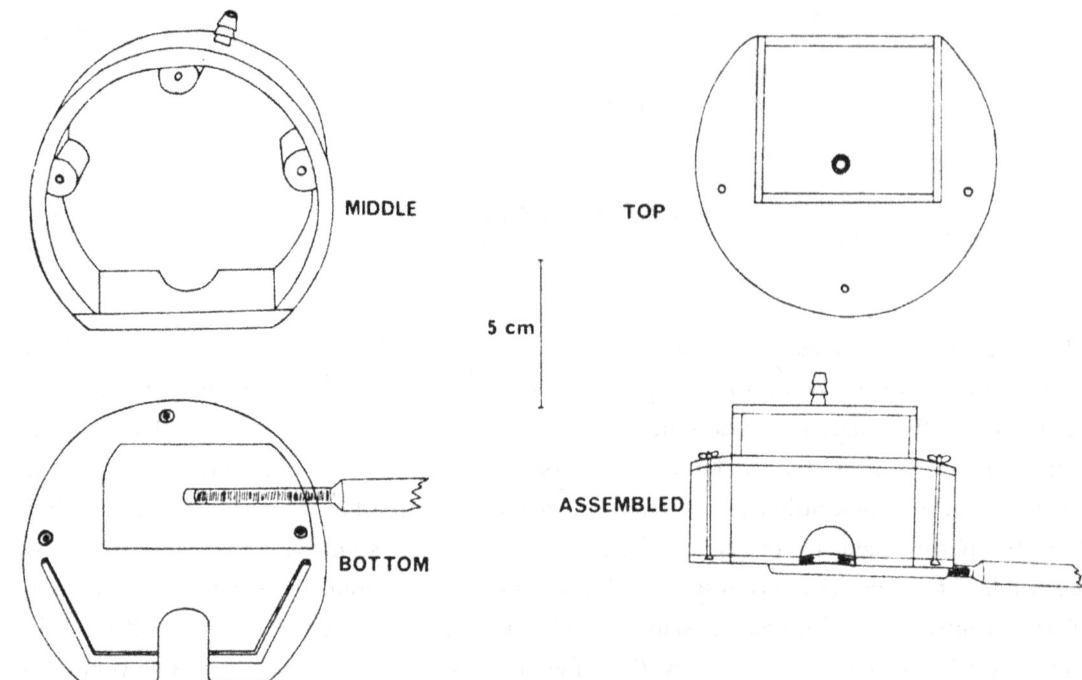

FIGURE 4.1: The three segments of the Plexiglas plethysmograph and the assembled unit held together with wing nuts on three bolts attached to the bottom plate and passing through the upper plates. Plastibase was applied as a sealant between each plate and liberally applied at the opening through which pass the hepatic artery, portal and hepatic veins, nerves, lymphatics, and bile duct. Reproduced with permission from Lautt WW, Brown LC, Durham JS. Active and passive control of hepatic blood volume responses to hemorrhage at normal and raised hepatic venous pressure in cats. *Can J Physiol Pharmacol* 58(9): pp. 1049–1057, 1980. © 2008 NRC Canada or its licensors. (This figure from publication Can J Physiol Pharmacol is reproduced with permission from publisher NRC Canada).

of the liver to the diaphragm were ligated and cut. The side wall was then slid over retaining bolts and a plasticized hydrocarbon gel was used to seal the junction. An aperture in both the wall and lower plate allowed unimpeded passage of the attachments of the liver to the posterior abdominal wall and the blood vessels and nerves to and from the liver. The portal vein and hepatic artery remained intact. A Plexiglas lid with a fluid outlet was fastened to the top wall and sealed with the gel. The plethysmograph was then filled with warm Ringer's solution and the outlet from the plethysmograph was connected to a float recorder operating an isotonic transducer. This procedure can be coupled with electrical stimulation of the hepatic nerves and recording of blood flow from the hepatic artery. Portal pressure can be measured and is an excellent index of proper placement of the plethysmograph. Small elevations in central venous pressure can be shown to reflect through the liver to the portal vein with an accompanying rapid increase in hepatic blood volume.

In the first report of the use of the hepatic plethysmograph, we wanted to mimic the procedures that the Swedes had so successfully used for studying capacitance responses in skeletal muscle and intestine. They had access to the venous outflow from these tissues and could therefore obtain appropriate samples of blood and measure total blood flow as well as control the venous blood pressure. The problem we faced with the liver was much more complex because of the unique hepatic vascular anatomy. The hepatic veins drain directly into the inferior vena cava just beneath the diaphragm and cannot, therefore, be directly isolated.

The approach used was referred to as the hepatic venous long-circuit (Figure 4.2). Vena caval blood flow draining toward the heart from the lower body was diverted and drained retrograde through femoral venous catheters. The vena cava was ligated between the adrenal veins and the hepatic veins so that the only flow remaining in the thoracic vena cava was entirely from the liver. The thoracic vena cava was cannulated with a molded glass catheter and drained from the chest cavity into an extracorporeal blood reservoir, which also received the flow from the lower vena cava. Thus, all of the venous return from the inferior vena cava was collected through two outlets that could be precisely calibrated simply by timing the effluent in a graduated cylinder, with continuous measurement using flow probes. The hepatic venous outflow allowed complete control over hepatic venous pressure and afforded samples of pure mixed hepatic blood. Blood in the extracorporeal reservoir is continuously monitored for volume and the warmed blood is pumped back to the cardiac venous return through catheters in the jugular vein. The successful application of the plethysmograph and blood flow measurements to the liver allowed protocol designs and data interpretation to be "fast tracked" because of the very clear conceptual guidance provided by the Swedish scientists, Folkow, Mellander, Johansson, Kjellmer, and others.

In that first study [123], the ability to precisely manipulate and quantify factors affecting hepatic blood volume were demonstrated with a clarity that fostered my appreciation of the necessity of having appropriate tools to precisely quantify biological responses in vivo. We determined the capacitance responses to electrical nerve stimulation and demonstrated a hyperbolic relationship that showed a maximal response expelling approximately 50% of the total hepatic blood volume at a nerve stimulation frequency of 6–8 Hz. A 2-Hz stimulation produced approximately 50% of the maximal response. The ability to produce nerve frequency stimulation/capacitance response curves as well as vasoactive dose/response curves allowed for the application of classic pharmacodynamic approaches to be used. Pharmacodynamics of vascular smooth muscle was previously developed in vitro, using organ bath techniques of a wide range of sophistication. However, extrapolation of these concepts to intact vascular beds and the gross responses that are quantified in an entire vascular bed generally had not been done.

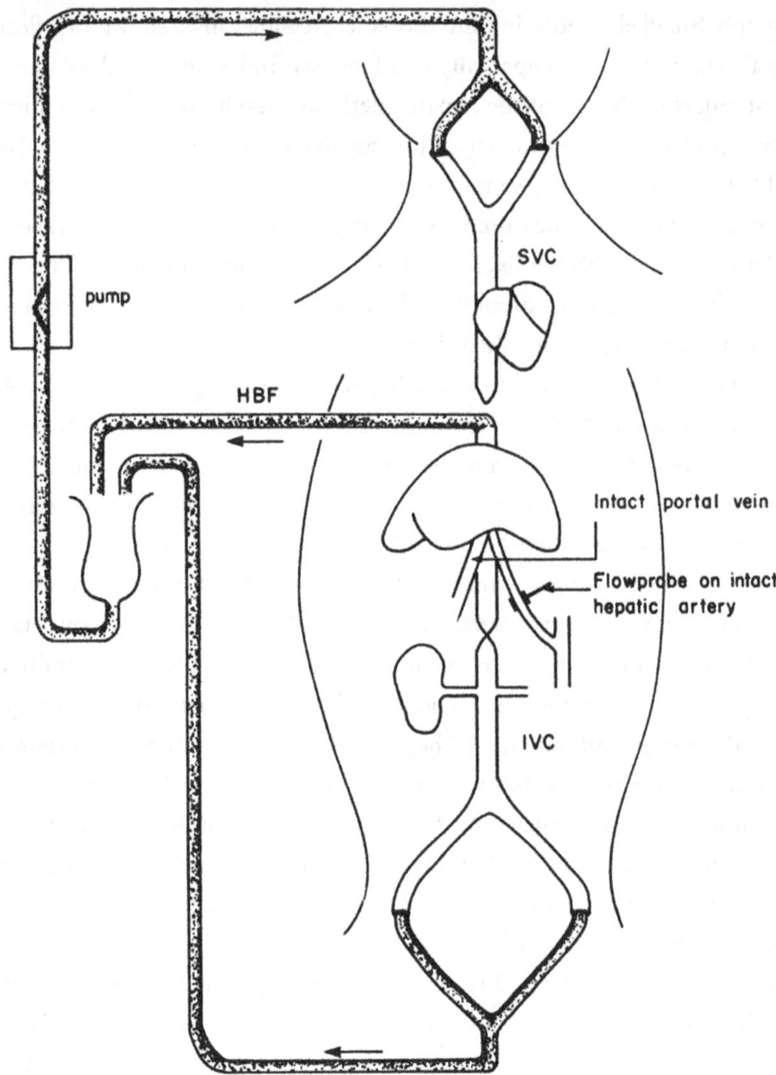

FIGURE 4.2: Hepatic venous long-circuit preparation with hepatic arterial flow (HAF) recording. Hepatic venous cannula allows measurement of total hepatic blood flow (HBF), control of hepatic venous pressure, and sampling of pure mixed hepatic venous blood. Portal flow is calculated from HBF – HAF. This preparation allows measurement of hepatic hemodynamics and uptake or output of substances quantitated in blood. Reproduced from Lautt WW. Method for measuring hepatic uptake of oxygen and other blood-borne substances in situ. *J Appl Physiol* 40(2): pp. 269–274, 1976. (This figure from publication J Appl Physiol is reproduced with permission from publisher).

4.1 HEPATIC BLOOD VOLUME

For an organ to serve as a physiological blood reservoir, it must contain blood that can be mobilized without negative consequences to the organ. The liver is a major blood reservoir with approximately 25–30% of the liver volume accounted for by blood [114]. Fifty to sixty percent of this blood volume can be expelled by sympathetic nerve action within 90 s without impairment of liver function. The liver contracts in direct response to α_2-adrenergic and angiotensin agonists. Passive collapse due to reduced blood flow is dramatic in the liver [20,121,204].

Capacitance, or total blood volume, consists of stressed volume and unstressed volume [104,112,121,325,326]. Stressed volume depends on the relationship between intrahepatic pressure and hepatic compliance. Compliance is a measure of the elasticity of the vascular bed and is defined as the change in volume per unit change in pressure (many studies incorrectly interchange capacitance and compliance). Hepatic compliance is linear over physiological pressures (2.5–3 ml/mmHg per 100 g or 0.6 ml/mmHg per kilogram of body weight) [121]. If this linear relationship between intrahepatic pressure and volume is extrapolated to zero pressure, the intercept is positive, that is, a theoretical volume remains in the liver at zero venous pressure (Figure 4.3). This extrapolated volume is the unstressed volume. These concepts have major physiological importance. The unstressed volume accounts for approximately 40% of the total hepatic volume [121]. The unstressed volume is hemodynamically inactive in that relationships between pressure, flow, and volume would not be altered if this blood pool did not exist. However, it is the unstressed volume that is actively regulated.

Active venoconstriction occurs mainly, if not entirely, by changing the unstressed volume [112], that is, compliance is not changed. The effect of norepinephrine on liver volume is to convert unstressed volume to stressed volume without altering compliance [121]. Active contraction of the capacitance vessels transfers blood from the "inactive" pool of unstressed volume to the "active" pool of stressed volume, which maintains venous pressure and venous return. Passive blood mobilization of stressed volume is secondary to reduced intrahepatic pressure, and the volume of blood is expelled according to the compliance of the vascular bed.

Although both active and passive regulation can occur, they operate via the two separate pools and have quite different consequences (also see Figure 14.1 in Chapter 14). Determining the extent of active and passive regulation requires determination of changes in stressed volume and unstressed volume, hepatic compliance, and intrahepatic pressures. Two examples, previously discussed in more depth [114], will illustrate the complex interactions. The response to sympathetic nerves and norepinephrine involves reduced hepatic blood flow. The passive effect of reduced blood flow is to produce a linearly related decrease in intrahepatic pressure and therefore in stressed volume. However, to conclude that a major component of the volume response to norepinephrine is secondary to the passive effects of flow reduction is not correct. When flow is reduced by mechanical

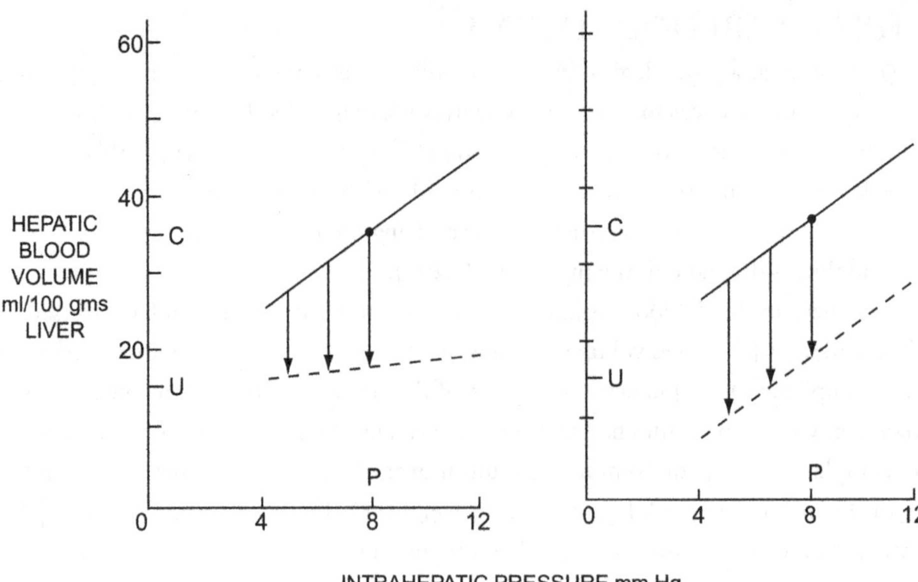

FIGURE 4.3: The pressure–volume relationship in the hepatic venous bed (solid line). Hepatic blood volume, or capacitance (C), at normal portal venous pressure is made up of unstressed volume (U) plus the stressed volume. The slope of the relationship (solid line) is the compliance. The stressed volume is determined by the compliance and the transmural pressure (P). Venoconstriction, shown by the arrows, could result from a change in compliance with no change in unstressed volume (broken line, left panel), a change in unstressed volume with no change in compliance (broken line, right panel), or a combination of changes in both compliance and unstressed volume. Active venoconstriction affects unstressed volume as compliance does not appear to change in response to the major constrictors. The biological response is seen in the right panel. Reproduced with permission from Greenway CV, Lautt WW. Blood volume, the venous system, preload, and cardiac output. *Can J Physiol Pharmacol* 64(4): pp. 383–387, 1986. © 2008 NRC Canada or its licensors. (This figure from publication Can J Physiol Pharmacol is reproduced with permission from publisher NRC Canada).

reduction of hepatic inflow, intrahepatic pressure falls, and passive reduction in hepatic blood volume occurs (stressed volume). However, during sympathetic nerve stimulation, intrahepatic pressure rises because of constriction of the hepatic venous sphincters so that the passive consequence of the intrahepatic pressure would be to expand stressed volume, not decrease it. The net effect is a decreased total volume and a decreased unstressed volume but an increased stressed volume. If flow and pressure also decreased, the extra stressed volume can be expelled as well (Figure 4.4).

During the response to hemorrhage, active mechanisms will reduce liver volume by reducing unstressed volume; venous return, cardiac output, hepatic blood flow, and intrahepatic pressures will be protected. If, however, all of the putative active regulators are removed (adrenalectomy, he-

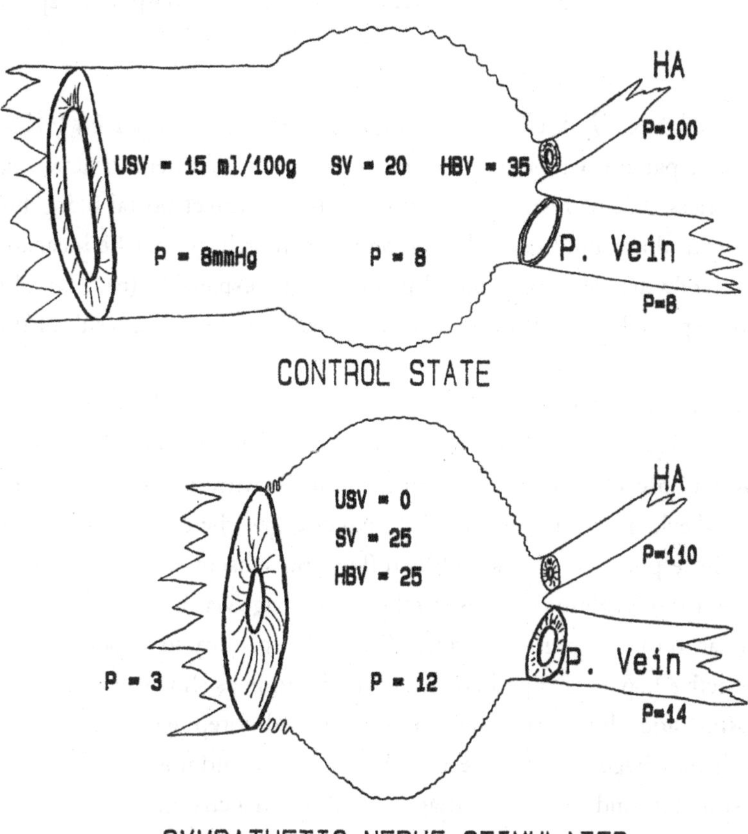

FIGURE 4.4: Unstressed volume (USV) is dramatically reduced. The compliance is probably not altered (see text) but the hepatic venous sphincters constrict, thus raising pressure (P) in the sinusoids and elevating stressed volume (SV). Total hepatic blood volume (HBV) is decreased. In the basal state, the presinusoidal resistance is trivial and no significant pressure gradient exists between the portal vein and sinusoids; nerve stimulation constricts this site and creates a gradient. Reproduced from Lautt WW. Hepatic circulation. In: *Nervous Control of Blood Vessels.* Harwood Academic Publishers GmbH, London, UK. Chapter 13, Figure 13.7, p. 487, 1996. (This figure from publication Nervous Control of Blood Vessels is reproduced with permission of publisher Harwood Academic Publishers GmbH).

patic denervation, nephrectomy, and hypophysectomy), the effect of blood loss is more profound, the reduced cardiac output leads to greater decreases in flow and pressures, the effect on hepatic stressed volume is increased. Whereas the blood mobilized by unstressed volume is reduced, the contribution of stressed volume is increased. The liver still compensates for approximately 21% of the hemorrhaged volume [204]. Thus, although the passive effect of reduced liver flow is large, without knowing the intrahepatic pressure, one cannot assess the degree of volume change that is active (unstressed volume) or passive, secondary to recoil of the compliant hepatic vascular bed (stressed volume).

4.2 CAPACITANCE FUNCTIONS IN DISEASED LIVERS

The severely cirrhotic patient will perish most likely as a result of a hemorrhagic episode resulting from rupture of varices that represent portacaval shunts that direct portal blood flow around the liver. The major organ that would normally respond to hemorrhage may be incapable of adequate response in the severely diseased state. Rapid blood volume expansion (e.g., liquid consumption) might similarly be expected to result in cardiovascular perturbations as a result of the less efficient volume-buffering capacity of the diseased liver.

In the 14-day bile duct ligated model of liver disease in cats, the blood volume compensation for hemorrhage was impaired [335]. The liver compensated for 20% of the blood loss in sham-operated animals but only 11% of blood loss in the animals with diseased livers. Surprisingly, despite the histological evidence of hepatic architectural changes, the compliance calculated from the decreased volume in response to decreased blood flow and subsequently decreased portal pressure was unaltered. The stressed volume response to the hemorrhage was well maintained in the diseased liver and was similar to that seen in the healthy livers. The inability to respond to hemorrhage was secondary to a selective hepatic sympathetic nerve dysfunction as shown by the lack of response to direct electrical stimulation but normal responses to infused norepinephrine. The healthy animals responded to the hemorrhage with a decrease in both stressed and unstressed volume, whereas the cats with liver disease responded to hemorrhage only through a change in the stressed volume.

The maintenance of normal capacitance responses to norepinephrine but impaired responses to sympathetic nerve stimulation was unexpected because chemical denervation (6-hydroxydopamine)-induced sympathectomy of the liver [206] leads to a denervation supersensitivity where capacitance responses to norepinephrine were modestly increased over the response in the healthy liver. Of note was that the response to portal pressure was dramatically increased, showing very significant denervation supersensitivity. This increase in portal and intrahepatic pressure would have increased the stressed volume of the liver, thereby underestimating the effect of norepinephrine on the unstressed volume. The sympathetic nerve dysfunction in the diseased liver model appeared to be restricted to the liver as systemic vascular pressor responses to bilateral carotid occlusion were well maintained.

• • • •

CHAPTER 5
Resistance in the Hepatic Artery

5.1 INTRINSIC BLOOD FLOW REGULATION: THE HEPATIC ARTERIAL BUFFER RESPONSE AND AUTOREGULATION

The liver does not control portal blood flow, which is simply the outflow of the extrahepatic splanchnic organs. If the vascular resistance to portal flow is increased to a maximum level by, for example, stimulation of the hepatic sympathetic nerves, portal pressure rises but portal blood flow does not fall. Portal blood flow is determined by the net outflows of the splanchnic organs including the stomach, spleen, pancreas, intestines, and omentum. If portal blood flow changes, the hepatic arterial flow changes in the opposite direction, thus tending to maintain total hepatic blood flow constant. This mechanism does not completely compensate but rather buffers the effect of portal blood flow changes on total hepatic flow, thus accounting for the name I chose for this response, the hepatic arterial buffer response (HABR) [192].

The HABR is proposed to operate by the following mechanism (Figure 5.1). Adenosine is constantly secreted into the space of Mall, which is a very small isolated fluid compartment through which passes the portal triad, consisting of the fine terminal branches of the hepatic artery, portal vein, and bile ductule, all in intimate contact. The concentration of adenosine, a potent vasodilator, is regulated by the rate of washout from the space of Mall into the blood vessels. When portal blood flow decreases, less adenosine is washed away, and the elevated adenosine concentration leads to dilation of the hepatic artery. This is the suggested mechanism of the HABR.

A second form of intrinsic hepatic arterial regulation is referred to as arterial autoregulation, which was previously believed to be myogenic in nature. If arterial pressure is decreased, this leads to a reduction in arterial blood flow, which washes away less adenosine from the space of Mall. The accumulated adenosine leads to arterial dilation. The HABR and autoregulation work through the same mechanism and operate simultaneously, thus leading to potential complex interactions. In addition, extrinsic factors such as circulating hormones, autonomic nerves, and various nutrients can interact to affect the hepatic circulation in a number of complex ways.

5.1.1 Metabolism and Hepatic Blood Flow

The first clue that hepatic metabolic demands did not control hepatic arterial flow came from a serendipitous observation made at the end of experiments designed for other purposes. In these

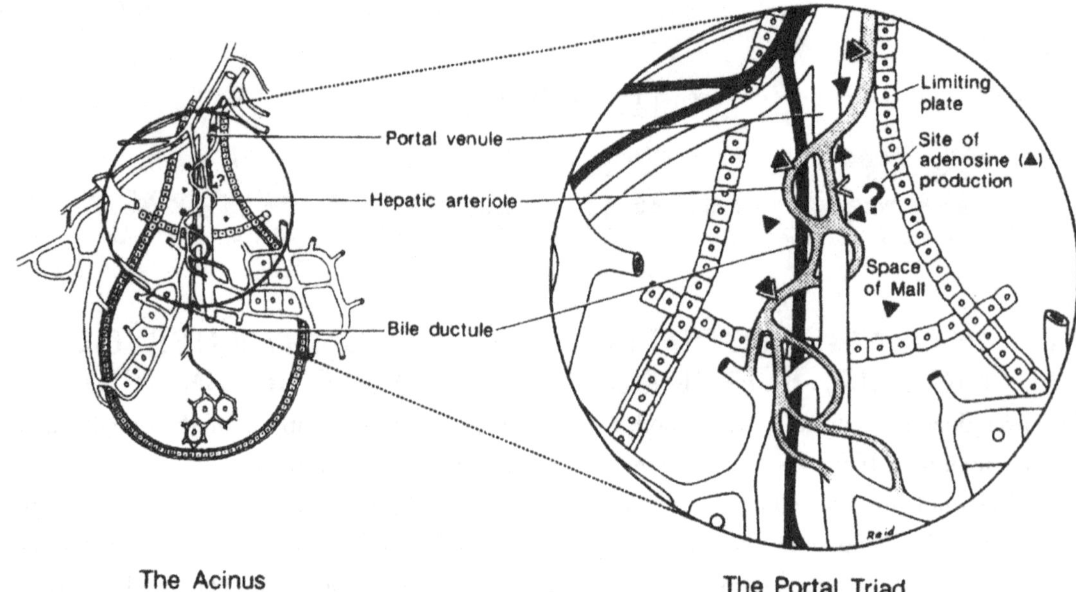

FIGURE 5.1: Adenosine washout hypothesis for the integrated intrinsic flow regulators of the hepatic artery: the HABR and autoregulation. The hepatic acinus represents a cluster of parenchymal cells arranged like a 2-mm berry on a vascular stalk, the portal triad, which leads into the center of each acinus. In contrast with the schematic figure, clear delineation does not exist between adjacent acini, the outer zones of adjacent acini possibly draining into more than one hepatic venule. The terminal branches of the hepatic arteriole, portal venule, and bile ductule lie within the enclosed space of the portal triad delimited by a limiting plate of cells. The fluid surrounding these vessels and contained within the limiting plate comprises the space of Mall. Adenosine is proposed to be continuously secreted into the space of Mall (independent of general parenchymal cell oxygen supply or demand), and the local concentration of adenosine determines the tone of the hepatic artery. Adenosine can be washed away into the bloodstream of the portal vein or hepatic artery. The HABR is accounted for by this mechanism, whereby a reduced portal blood flow washes away less adenosine and the accumulated adenosine leads to dilation of the hepatic artery. Classic autoregulation is explained by the same adenosine washout hypothesis. For example, increased arterial blood pressure leads to increased arterial blood flow with a subsequent washout of adenosine and resultant arterial constriction. The unique anatomical arrangement precludes substances released from the parenchymal cells from diffusing upstream against the blood flow to affect hepatic arterial tone. Reproduced with permission from Lautt WW. The 1995 Ciba-Geigy Award Lecture. Intrinsic regulation of hepatic blood flow. *Can J Physiol Pharmacol* 74(3): pp. 223–233, 1996. © 2008 NRC Canada or its licensors. (This figure from publication Can J Physiol Pharmacol is reproduced with permission from publisher NRC Canada).

experiments, the hepatic circulation was studied using the hepatic venous long-circuit protocol, draining blood from the liver into a reservoir, from which it was pumped back to the animal through the jugular veins (Figure 4.2, Chapter 4). The lower vena cava (below the hepatic veins) drained retrograde via two femoral venous catheters into the same warmed reservoir. In this way, we could measure total hepatic blood flow, control hepatic venous pressure, and measure liver metabolism in a preparation that had intact nerves and inflow vessels [184]. At the end of experiments, which were designed to study metabolic effects of the hepatic nerves, the blood reservoir would become depleted at a time when the biological preparation was still functional. At the point where the reservoir was almost empty, we added a solution of Dextran 75 and Ringer's solution to produce an isovolemic hemodilution. We anticipated that the oxygen delivery to the liver would decrease and that the hepatic artery would subsequently dilate. Although the oxygen delivery decreased to 68% of control and hepatic venous and portal oxygen content declined, the hepatic artery did not show a dilation. In fact, it showed a small constriction, and total hepatic blood flow remained constant. The results were completely unexpected and quite disconcerting, and our first fear was that the surgical preparation we had used had produced an abnormal vasculature that was incapable of responding to a decreased oxygen supply. We quickly proved that the artery was capable of vasodilation by demonstrating a brisk response to isoproterenol. These procedures were repeated at the end of every experiment in several series so that we eventually had 23 animals that had been subjected to a large isovolemic hemodilution with full hepatic arterial, portal venous, hepatic venous, and systemic hemodynamic and blood gas data monitored. Before the availability of personal computers, we attempted to determine factors that correlated with changes in hepatic arterial blood flow, and after 150 hand-graphed correlations, we determined that the only unexplained correlations were with portal venous flow and hepatic arterial conductance. It appeared that in situations where the portal flow increased, hepatic arterial conductance decreased, and when portal flow decreased, hepatic arterial conductance increased. The end result was that changes in portal flow tended to produce opposite changes in hepatic arterial flow such that total hepatic blood flow was maintained at a remarkably constant level. This work had two surprising conclusions. The first was that the oxygen supply-to-demand ratio for the liver did not appear to be an appropriate stimulus to change hepatic arterial blood flow. The second was that the major factor that appeared to be controlling hepatic arterial blood flow was portal venous flow [185].

We continued examination of this first observation using dinitrophenol to increase oxygen demand or SKF525A to inhibit metabolism. The hepatic artery showed no tendency to change in response to the metabolic alterations; the hepatic oxygen demands were maintained solely by altered hepatic extraction [191]. Scattered reports had been consistent with this conclusion. As early as 1950, Myers et al. [285] reported that an increased splanchnic metabolism coexistent with a normal splanchnic blood flow provided an exception to the hypothesis that the rate of local tissue metabolism regulates the blood flow through the liver. Chronic alcohol exposure in rats, for example, led to elevated oxygen demand by the liver, but the hepatic artery actually constricted [33]. Lactic acidosis in dogs resulted in elevated portal flow but reduced oxygen delivery to the liver and reduced hepatic venous oxygen content. The hepatic artery did not respond to the decreased oxygen supply, but rather it constricted in response to the elevated portal flow [139].

Studies relating enzyme induction to blood flow and the effect of bile salts had suggested hepatic metabolic regulation of blood flow. Ohnhaus et al. [293] reported that enzyme induction with phenobarbitone increased hepatic blood flow in the rat, an observation that was generally taken to support the notion that blood flow in the liver was metabolically controlled. However, Nies et al. [287] reported that, in rats, hepatic enzyme induction with 3,4-benzpyrene and 3-methylcholanthrene caused no hemodynamic changes. In contrast, phenobarbital induction did cause elevated hepatic blood flow, but the augmented flow was via the portal vein, indicating that the increased hepatic flow was a result of an extrahepatic phenomenon unique to phenobarbital and unrelated to induction of hepatic enzymes.

In examining the effect of stimuli on metabolic activity and vascular responses, it is important to differentiate direct from indirect effects. One response that initially appeared to contradict the absence of hepatic metabolic control of arterial flow was the observation that bile salts both stimulate hepatic metabolism and cause hepatic arterial dilation. The vascular and metabolic responses of the liver to bile salts are, however, independent and occur at different doses [209]. Low doses of taurocholate (1 μM/min per kilogram of body weight) infused into the portal vein produced elevated bile flow but no arterial vasodilation, whereas higher doses produced dose-related vasodilator responses in both the superior mesenteric and hepatic arteries. The independence of vascular and metabolic effects was more clearly shown when the same dose of taurocholate infused into the hepatic artery or portal vein produced equal effects on bile flow but considerably greater arterial responses when infused into the artery. The bile salts thus produce direct arterial vaodilation unrelated to hepatic metabolic stimulation. The hepatic artery is not subservient to hepatic metabolic activity [195].

5.1.2 Portal Flow Regulation of Hepatic Arterial Flow

The earliest studies reporting an effect of changes in portal perfusion on hepatic arterial flow were credited by Child (1954) to Betz in 1863 and Gad in 1873 [87]. The observation was studied periodically by very few people over the next 70 years. Various hypotheses were proposed to account for this interaction, and it was generally concluded that a myogenic mechanism accounted for what became referred to as the reciprocal flow relationship of the hepatic artery and portal vein. This was the state of affairs in the mid-1970s and was matched with the strong statement in textbooks (based on no experimental evidence) that the hepatic artery was under metabolic control of the liver. With the discovery that the hepatic artery was not under metabolic control, we focused on determining the mechanism of vascular interactions. Very quickly we determined that there was no reciprocal relationship between the blood flows and that changes in hepatic arterial flow did not change portal venous flow or portal resistance.

The decrease in portal pressure that may be seen with complete occlusion of hepatic arterial inflow is not caused by a decreased vascular resistance of the portal vessels but rather is caused by the fact that the intrahepatic and portal pressure is largely maintained as a result of blood flowing through postsinusoidal resistance vessels located in the small hepatic veins, so that a decrease in total flow subsequent to arterial occlusion results in a reduced pressure.

Because the relationship between the portal vein and hepatic artery was clearly not a reciprocal flow relationship, we coined the term *hepatic arterial buffer response* (HABR) [192] to acknowledge the role of the hepatic artery in buffering changes in total hepatic blood flow that would occur secondary to changes in portal flow. The primary function of the hepatic artery was recognized as being maintenance of hepatic blood flow, per se, regardless of oxygen supply or demand.

We considered a number of alternate hypotheses to account for the mechanism. Mechanisms we considered and rejected have been reviewed [192]. They included a myogenic mechanism, which proposed that a change in portal flow would produce a change in portal pressure that would be sensed by the hepatic artery. Changes in portal pressure are very minor even in the face of very large changes in portal flow. We also considered a neural mechanism, but we and others have shown that the HABR is seen in a fully denervated liver [257,331] and in transplanted human livers [131,299].

We considered the possibility of purely physical interactions that would result from the mechanical consequences of interposing a slower flowing stream (the portal blood) into the path of a faster flowing stream (the hepatic artery). The absence of a HABR in the isolated liver was the principal evidence to preclude this purely physical model. The physical model also predicted that occlusion of the hepatic artery should result in a reduction in portal conductance, which was not seen.

We also considered that the quality of the portal blood was an important factor. If portal blood contained a constrictor agent, an elevation in portal flow might deliver more constrictors to the hepatic vasculature and result in arterial constriction. We eliminated this possibility by devising a preparation that had a dual vascular shunt such that the portal outflow from the guts could be sent either to the liver or to the vena cava. A similar shunt from the vena cava allowed us to perfuse the liver with vena caval blood. Switching from portal blood to vena caval blood perfusing the liver did not result in any change in the hepatic arterial conductance and the buffer capacity was similar regardless of the source of portal flow.

We considered metabolic control, whereby metabolic by-products or oxygen content were a factor. Our earlier studies had indicated that oxygen was not a regulatory factor, and we only later realized that the vascular anatomy of the hepatic acinus precluded metabolic feedback to the hepatic arterial resistance vessels.

We were finally left with the remaining hypothesis, which we referred to as the washout hypothesis. I proposed that a dilator substance is produced within the space of Mall at a constant rate and that the concentration is dependent on the rate of washout into the portal blood (Figure 5.1). According to this theory, a decrease in portal flow would wash away less of the dilator, thus leading to an accumulation and arterial vasodilation. The challenge then was to identify the dilator. In considering possible dilators, we formulated a list of more than 20 candidates. The frustration felt at this daunting list was exemplified by a piece of doggerel that I wrote at that time [192]:

Lautt's Lament
I think that I shall never know
the factor controlling hepatic blood flow.
In spite of our tests and our pontification,
we still don't know what causes dilation
of the hepatic artery, which stands at the ready
keeping total blood flow to the liver quite steady.

When the portal blood flow is quickly reduced
the artery dilates, or so we've deduced,
as result of accumulated dilator
that relaxes smooth muscle and makes the flow greater.
But what is this magical dilator stuff?
We don't know yet for we've not done enough.

For a compound to be eligible as the dilator substance controlling the HABR via a washout mechanism, several criteria must be met. These include, but are not restricted to, the following: (i) the putative regulator must dilate the hepatic artery; (ii) portal blood must have access to the arterial resistance vessels so that portal flow can wash away the substance from the area of the resistance vessels; (iii) potentiators of effects of the putative regulator should also potentiate the buffer response; (iv) blockers of effects of the putative regulator should also inhibit the buffer response.

In a preliminary screening approach, based on only one to three tests per compound, the buffer response was unaltered by prior administration of atropine, propranolol, ouabain, aminophylline, theophylline, indomethacin, metiamide, and mepyramine. At that time I was fortunate to have John W. Phillis as my Department Head. He was involved with studies related to the functions of adenosine in the central nervous system and had access to some new pharmacological tools. His suggestion that adenosine was the "magical dilator stuff" initially was received with skepticism because of our previous observation that the hepatic artery was not affected by the metabolic status of the liver. At that time it was generally regarded that adenosine served as a major link between tissue metabolism and blood flow. Furthermore, aminophylline and theophylline had not blocked the HABR. The first compounds tested were isobutylmethylxanthine, as an adenosine receptor antagonist, and dipyridamole, as an adenosine uptake blocker. Serendipity again played a hand, and we showed significant blockade with isobutylmethylxanthine and potentiation with dipyridamole. Later studies indicated that these compounds had many other actions and that we were fortunate indeed to have been able to demonstrate the appropriate effects with them [220]. We were sufficiently cautious to delay publication until we had confirmed the data, using more selective compounds. Nevertheless, those experiments allowed us to fulfill the important criteria discussed above. Adenosine was shown to produce a powerful dilation of the hepatic artery. Portal flow was shown to have access to the arterial resistance vessels by the observation that intraportal adenosine produced a profound dilation of the hepatic artery. Dipyridamole potentiated the effect of infused adenosine and potentiated the buffer response. Isobutylmethylxanthine blocked the HABR and blocked the response to exogenous adenosine.

To study the buffer response, we had to develop a surgical protocol that allowed repeated buffer responses to be quantified. A standard preparation was used to do mechanistic studies of the buffer response (Figure 8.5, Chapter 8). Using this system, all of the portal blood flow is supplied by the superior mesenteric artery, which allows measurement and control of portal flow at the arterial inflow side. Hepatic arterial and superior mesenteric arterial blood flow are measured using flow probes. An occlusion of the portal inflow results in a rapid dilation of the hepatic artery (Figure 5.2, left panel). At that point it is important to control hepatic arterial blood pressure at a constant level so that any change in arterial flow is secondary only to the buffer response and not to changes in systemic arterial pressure.

The availability of good pharmacological tools is often a primary limitation in physiological studies. With the development of 8-phenyltheophylline, another gift from J.W. Phillis, we had access to a much improved adenosine receptor antagonist. This compound is certainly not perfect because it has to be administered in a very alkaline solution and is poorly soluble. Nevertheless, it has proven to be a useful tool (Figure 5.3) that quantified dose-related competitive inhibition of exogenous adenosine and parallel inhibition of the buffer response [216]. The inhibition of the response to adenosine and the buffer response are shown from one cat (Figure 5.2).

5.1.3 Autoregulation

It occurred to us that if adenosine was being washed away from the space of Mall by portal blood flow, it should equally be able to be washed away by arterial flow. We thus suggested that what was previously considered to be a myogenic mechanism might, in fact, be mediated by the adenosine washout

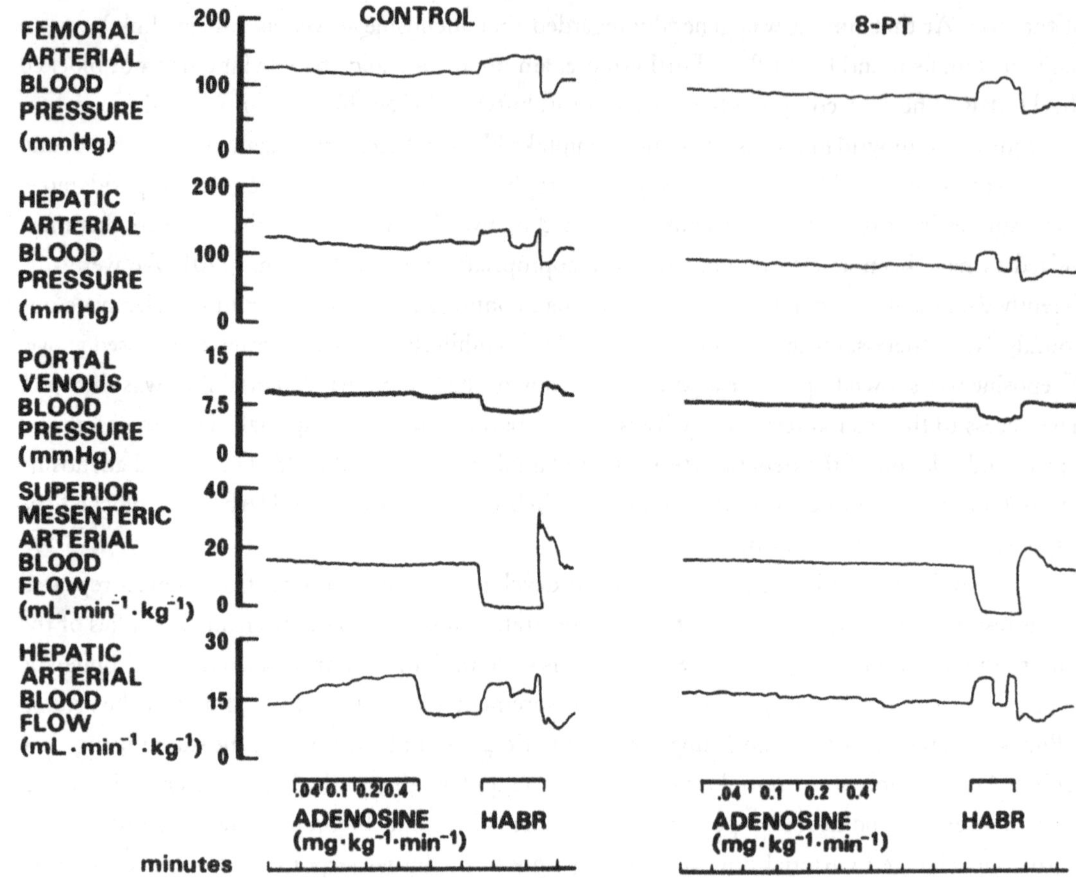

FIGURE 5.2: The response to adenosine and the HABR and blockade by adenosine receptor antagonist. Stepwise dose–response relationship for intraportal infusion of adenosine followed by the HABR elicited by complete occlusion of portal inflow vessels. Note that during HABR, the response is measured after hepatic arterial blood pressure has been adjusted back to the control level, using a vascular occluder, in order that all the increase in hepatic arterial flow is due to vasodilation by the HABR mechanism. The left panel indicates that adenosine in the portal vein has access to the arterial resistance vessels. The right panel shows complete blockade of vasodilator response to exogenous adenosine and to the buffer response by the competitive adenosine receptor antagonist 8-phenyltheophylline (8-PT). Dilator effects of isoproterenol were not affected in these studies. Note the potential artifact if hepatic arterial pressure is not controlled, as in the right panel. Reproduced with permission from Lautt WW, Legare DJ. The use of 8-phenyltheophylline as a competitive antagonist of adenosine and an inhibitor of the intrinsic regulatory mechanism of the hepatic artery. *Can J Physiol Pharmacol* 63(6): pp. 717–722, 1985. © 2008 NRC Canada or its licensors. (This figure from publication Can J Physiol Pharmacol is reproduced with permission from publisher NRC Canada).

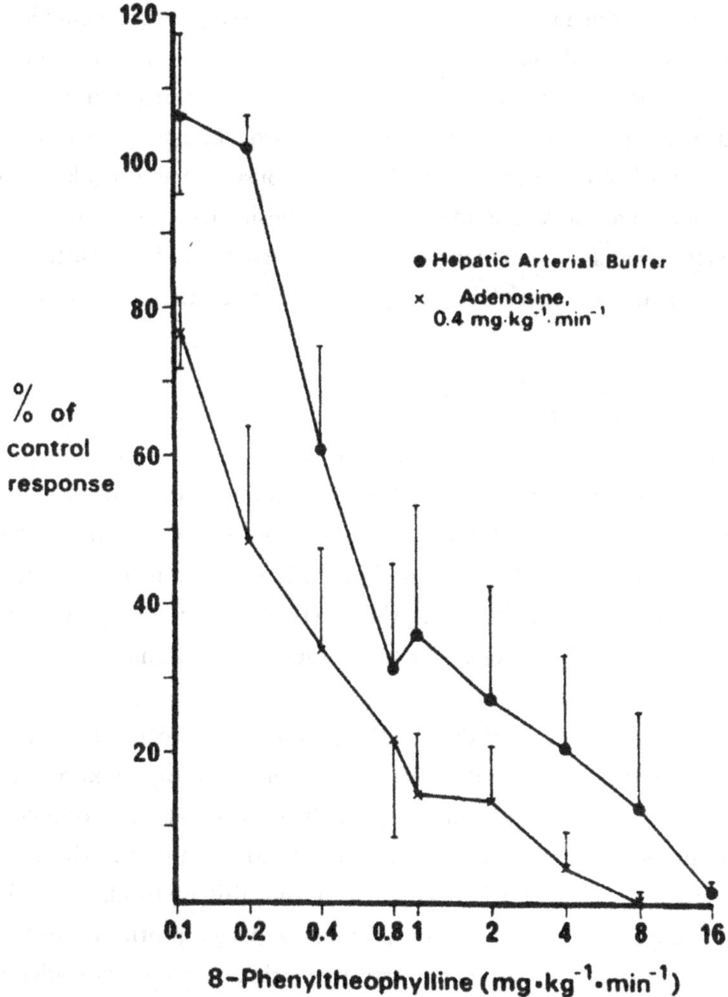

FIGURE 5.3: The progressive competitive antagonism of the hepatic arterial vasodilator effect of complete reduction of portal blood flow (the hepatic arterial buffer response) and the dilator effect of intraportal adenosine in the presence of progressive increases in intra-arterial doses of adenosine receptor antagonist (8-phenyltheophylline). The control dilation (percent change in conductance) seen with the buffer response and with the response to adenosine infusion was classified as 100%. The first responses plotted were in the presence of 0.1 mg/kg/min of antagonist. Note the parallel depression of vasodilator effect of exogenous adenosine and the buffer response. Roughly four times the dose of 8-phenyltheophylline is needed to produce equivalent depression of the buffer response compared with the response to exogenous adenosine. Reproduced with permission from Lautt WW. The 1995 Ciba-Geigy Award Lecture. Intrinsic regulation of hepatic blood flow. *Can J Physiol Pharmacol* 74(3): pp. 223–233, 1996. © 2008 NRC Canada or its licensors. (This figure from publication Can J Physiol Pharmacol is reproduced with permission from publisher NRC Canada).

mechanism. According to this hypothesis, an increase in arterial pressure would lead to an increase in arterial flow and a washout of the dilator. The resultant decrease in basal level of adenosine would lead to constriction of the artery. Conversely, a decrease in arterial perfusion pressure would lead to reduced flow and reduced washout of adenosine and subsequent arterial dilation. This hypothesis was confirmed, and we demonstrated that the buffer response and autoregulation were blocked in parallel with the response to exogenous adenosine by adenosine receptor antagonists [77].

The observation that autoregulation could be demonstrated to be consistent with the mechanism of the buffer response was crucial to our own acceptance of the adenosine washout hypothesis.

5.1.4 Quantitative Aspects of the HABR

If portal blood flow is severely reduced, the buffer response results in the hepatic artery dilating maximally as demonstrated by the inability to produce additional dilation in response to intra-arterial infusion of adenosine. Conversely, when portal flow is doubled, the hepatic artery constricts to a maximal extent, as demonstrated by the inability of intra-arterial norepinephrine to produce further constriction [222]. Thus, the HABR is sufficiently powerful to regulate the vascular tone in the hepatic artery over the full range from maximal vasodilation to maximal vasoconstriction.

To quantitate certain aspects of the HABR, a number of special techniques had to be developed. As previously discussed, it is important to be able to produce a known change in portal flow and to measure the change in hepatic arterial flow in the absence of a change in hepatic arterial perfusion pressure. We developed a method [220] to quantify the HABR for subsequent mechanistic studies (Figure 8.5, Chapter 8). Using this method, the entire portal blood flow is supplied through the superior mesenteric artery. Anastomotic connections to other arteries that were ligated (inferior mesenteric artery, gastric artery) provide adequate blood flow to those areas. The spleen is removed. This, unfortunately, leads to a reduced portal flow at the outset of the experiment, which results in a partially activated buffer response. This complication leads, in many experiments, to attempting to study the buffer response in an artery that is already nearly maximally dilated. Although such a method was necessary to carry out studies related to the mechanism of the HABR, the impact that this method has on the estimation of buffer capacity is unknown.

Buffer capacity is calculated as the change in hepatic arterial flow expressed as a percentage of the change in portal venous flow, where a 100% capacity would provide for full compensation. In anesthetized, splenectomized cats, the buffer capacity is only approximately 25% [220]. Others found a buffer capacity in the same range in anesthetized dogs [255] and pigs. Many studies report changes in portal and hepatic arterial flow but, without a steady hepatic arterial pressure, it is not appropriate to assume that calculated buffer capacity actually represents a true buffer response. This is well demonstrated by the following example (Figure 5.2). If perfusion pressure is not held constant, occlusion of the superior mesenteric artery produces a decrease in portal flow but an elevation of arterial pressure. The HABR tends to raise arterial flow, but autoregulation tends to counter this rise. After adenosine receptor blockade, both intrinsic mechanisms are blocked, so that the decrease in portal flow does not activate the HABR, but the rise in arterial pressure now causes a rise in hepatic arterial flow unimpeded by autoregulation [216]. Thus, without controlling for hepatic arterial perfusion pressure, one would conclude that adenosine receptor blockade had not affected the buffer response. With this precaution in mind, there are a number of publications that report compensatory changes in hepatic arterial flow that suggest a buffer capacity close to 100%. In our early studies, hemodilution [187] and metabolic inhibitors and stimulators [191] resulted in altered portal flow, with the hepatic arterial flow compensating to hold total hepatic blood flow within 4% of basal flow. In a cirrhotic rat model, without significant portacaval shunts, portal blood flow was significantly elevated and the hepatic arterial flow was reduced sufficiently to hold total hepatic blood flow at similar levels in the control and cirrhotic animals [79]. From the mean flow data, the buffer capacity was 72%. The constriction of the hepatic artery was unique in that all other arteries showed vasodilation, implying that the buffer overcame some general systemic dilator influence and yet produced an impressive compensation. Lactic acidosis in dogs [139] resulted in an increase in portal blood flow and a decrease in hepatic arterial flow resulting in a 100% compensation.

To carry out in vivo quantitative pharmacodynamic studies, it was also necessary to develop certain techniques. A key point is that changes in arterial vascular tone normally result in changes in flow, with minor changes occurring in systemic arterial pressure. In this situation it is imperative that changes in vascular tone be expressed as vascular conductance rather than resistance. Conductance is simply the inverse of resistance, and it is essential that the parameter that changes to the greater extent be in the numerator, that is, flow divided by pressure gradient rather than pressure gradient divided by flow [198]. One dramatic example will suffice for this chapter. Consider if a severe vasoconstriction is produced by sympathetic nerve stimulation or infusion of norepinephrine, with flow decreasing to approach zero. Conductance will also approach zero, but resistance will approach infinity. Clearly, resistance cannot be used to estimate an ED_{50} for drug effect. Resistance cannot be used even to calculate a simple arithmetic average [198] because resistance is nonlinearly related to flow. With the use of vascular conductance and nonlinear regression analysis, maximum responses (R_{max}) and the dose of drug that produces 50% of maximal vasoconstriction (ED_{50}), or the frequency of nerve stimulation that produces 50% of the maximal response (Hz_{50}), can be calculated. These parameters are important for determining the effects of pharmacological antagonists and modulators. Using this approach we were able to demonstrate that 8-phenyltheophylline is capable of blocking 100% of the buffer response. Furthermore, this approach allowed a clear demonstration that an exogenously administered receptor antagonist is capable of blocking exogenously administered agonists more effectively than the endogenous agonists. The ID_{50} for 8-phenyltheophylline to produce a 50% blockade of the effect of exogenous adenosine was 0.33 ± 0.03 mg/kg and the ID_{50} for blockade of the buffer was 1.31 ± 0.47 mg/kg. These data were recalculated from the data shown in Figure 5.3. Lack of recognition of the differential ability to block endogenous versus exogenous agonists can lead to serious error. For example, Mathie and Alexander [254] used one dose of 8-phenyltheophylline and demonstrated that that dose blocked the response to exogenous adenosine but did not completely block the buffer response. They concluded that factors other than adenosine must, therefore, be involved. Without demonstrating that a higher dose of blocker also left the buffer response incompletely blocked, this conclusion is inappropriate. It can be seen from Figure 5.3 that some doses of blocker could eliminate the response to adenosine but still leave a sizeable HABR.

Two other precautions must be considered that apply to in vivo pharmacology in general and to the use of adenosine receptor antagonists in particular. The first is that, because of the high variability in blocking effectiveness, we found it necessary to test for effective blockade in every animal and to double the blocking dose until full blockade of the response to infused adenosine is produced. The usual dose of 8-phenyltheophylline administered is 8 mg/kg. The second precaution relates to selectivity. As important as it is to show effective blockade, it must also be selective. That is, dilation to other agents, such as isoproterenol, should not be impaired.

Although the HABR has been shown in a wide variety of species, it has not been quantitated in terms of buffer capacity in conscious animals or humans. Clearly, to fully appreciate the homeostatic role of the HABR, it is essential to know the buffer capacity of the hepatic artery.

5.1.5 Roles of the HABR

If the hepatic arterial flow is not regulated by hepatic oxygen supply or demand, does this imply a serious pathophysiological problem for the liver with regard to hypoxic damage? The liver has often been referred to as an organ verging on hypoxia. This, in fact, is not the case. When oxygen delivery was reduced to the liver to 68% in our hemodilution studies, oxygen uptake was able to be maintained at completely normal levels simply by increasing the oxygen extraction from the available supply. Because of the unique microvascular anatomy, the liver is able to extremely efficiently extract compounds from the blood.

In many conditions, the HABR, despite being insensitive to reduced hepatic oxygen, serves to protect oxygen delivery coincidentally. For example, the hepatic arterial flow is preserved during hemorrhage. This relative sparing of the hepatic artery is secondary to the HABR responding to the decrease in portal blood flow [227]. Thus, the buffer response in this instance serves to preserve the oxygen supply of the liver without the oxygen supply being directly regulated.

In considering the teleological purpose for evolving a flow-regulatory system that tends to maintain blood flow, per se, constant to the liver, it was fortunate that we had recently concluded studies relating blood flow to hepatic drug clearance. The liver is involved with clearance of an extremely wide range of endogenous compounds including hormones, such as aldosterone and corticosterone [186,265], in a blood flow-dependent manner. It was proposed [186] that a major function of the liver might be considered to be an endocrine function according to the following logic. Traditionally, when one thinks of regulation of plasma hormone levels, only the endocrine gland that produces the hormone is considered. However, for the endocrine gland to be able to increase or decrease hormone levels in the blood, it is important that there be a reasonably rapid and quite constant clearance of the hormone to serve as a background against which secretion can lead to fine tuning. If hepatic blood flow was not prevented from rapid, transient changes secondary to similar changes in the portal venous flow, endocrine homeostasis would be imperiled. It thus seems highly likely that general endocrine and metabolic homeostasis is subserved by the function of the HABR.

A second consequence of a functional buffer response is related to the impact of altered hepatic blood volume on cardiovascular status. Hepatic blood volume is passively altered in response to changes in total hepatic blood flow [204]. The decrease in volume per unit decrease in blood flow is similar regardless of whether the reduction was in portal or arterial flow. Most resistance to portal blood flow through the liver is at the hepatic venous outflow [215]. Intrahepatic and portal pressures are altered by changes in blood flow through this resistance site. Changes in intrahepatic pressure lead to changes in hepatic stressed blood volume. Stressed volume is the product of compliance and intrahepatic pressure. The liver is extremely compliant and changes in intrahepatic pressure are also reflected back to the portal vein, which is also a very complaint system. The splanchnic venous system can account for two thirds of the blood volume response to hemorrhage and represents the largest blood volume reservoir in the body [112]. By tending to maintain hepatic blood flow constant and hence portal and intrahepatic pressures constant, transient alterations in hepatic blood volume and, therefore, in venous return are minimized by the HABR.

The HABR is capable of being activated at the acinar level and thereby plays a role in maintaining perfusion homogeneity and prevention of regional stagnation secondary to small local perturbations in regional pressure such as caused by external mechanical forces or shifts of abdominal contents. Maintenance of homogeneity at sinusoidal levels, smaller than the regions served by one terminal hepatic arteriole branch, are probably maintained through stellate cell activity (Chapter 2).

A further major homeostatic role for the adenosine washout mechanism is related to adenosine acting as a neurotransmitter. Reductions in portal flow, leading to elevated adenosine concentrations in the space of Mall, result in activation of hepatic sensory nerves, which leads to a reflex fluid retention and thus increased venous return. This mechanism is important for both physiological homeostasis and the pathology of salt and water retention in chronic liver disease (see Chapter 13).

5.1.6 Clinical Relevance

The existence of the HABR is reported to be the best prognostic indicator of a successful outcome for patient survival of a portacaval shunt to treat portal hypertension [37]. Similarly, it has been found that a brisk buffer response to a brief occlusion of portal flow is seen in patients after transplant [131] and is a useful indicator of suitable vascular reconnection.

That the liver cannot regulate its blood flow in accordance with the metabolic activity of the parenchymal cells is not disadvantageous to the liver under normal physiological conditions because of the excess oxygen delivered and the hepatic capacity to increase oxygen extraction. However, a relatively hypoxic liver shows increased toxic effects of alcohol [145], carbon tetrachloride [342], and halothane [262]. The formation of active toxic metabolites may occur to a greater extent and the ability of glutathione to detoxify such metabolites may be reduced when the NAD^+/NADH ratio is altered by hypoxia (Chapter 12). Although the HABR is seen in cirrhotic livers, the buffer capacity may be insufficient to maintain a normal oxygen supply [332], and there is suggestion that severely diseased livers lose the HABR [329].

The HABR shows considerable variability in diseased livers. The HABR is fully maintained in transplanted human livers [25,131] and appears to be maintained in liver diseases of considerable severity [11,127,152,282,316,330]. Portal flow remains high for as long as 2 years after liver transplantation, mainly as a result of elevated splenic blood flow, and is associated with reduced hepatic arterial flow [25].

If an HABR response to brief portal occlusion does not occur before establishment of a portacaval shunt to reduce portal hypertension, those patients will show the greatest reduction in portal pressure [389]. Unfortunately, those are the same patients with the poorest prognosis for survival [37]. The demonstration of an intact HABR, by observing an elevated portal and decreased arterial flow after a balanced liquid meal [64], has been suggested as a tool to assess the severity of liver disease because this response is decreased in very severely diseased livers [179].

5.1.7 Unresolved Issues

A number of aspects of the adenosine washout hypothesis have not been resolved or have not been tested. Although the anatomy is consistent with the space of Mall serving as the isolated fluid compartment into which adenosine is secreted to affect the hepatic artery and from which adenosine can be washed away by portal or arterial blood supply, no direct evidence exists to localize the anatomical site.

Another unresolved aspect of this hypothesis is the pathway of adenosine production. Hepatic arterial regulation is independent of liver metabolism. Adenosine production can occur by breakdown of adenine nucleotides or cyclic AMP; however, these sources are directly linked to the energy status of the cells. The liver generates adenosine from these sources, but the cells of production are parenchymal cells that are downstream from the hepatic arterial inflow resistance vessels. The unique microvascular anatomy of the hepatic acinus precludes metabolites released from the parenchymal cells from diffusing upstream to directly affect the hepatic artery. I suggest that the adenosine involved with the HABR and autoregulation is produced at a constant rate and secreted into the space of Mall and is most likely derived from demethylation of *S*-adenosylhomocysteine, a reaction that is oxygen independent and is proposed to account for basal adenosine production in the heart [243].

It is also unclear whether the rate of adenosine secretion can be modulated by some metabolic or hormonal responses that can lead to altered baseline secretion, thereby modulating the magnitude of the buffer response. In at least one third of animals tested under anesthesia and using the intensive surgical preparation required to study the mechanism of the HABR, the intrinsic hepatic arterial regulation and the response to exogenous adenosine are absent. We have eliminated several possible mechanisms for this interference with intrinsic regulation but currently cannot account for it. In a series of unpublished studies we demonstrated, to our satisfaction, that increasing the basal tone of the hepatic artery using infusion of vasoconstrictors does not return the HABR in animals where it is absent. Furthermore, we have not been able to link the absence to alterations in blood gases, pH, or lactic acid levels. Nitric oxide is also not involved with the HABR or autoregulation, in contrast to the superior mesenteric artery in which nitric oxide is shown to antagonize autoregulation [249]. It was our impression that in cats that had intestinal worms, or that had recently been treated for intestinal worms, the HABR is weak or absent. Furthermore, we have witnessed the HABR disappearing during an experimental protocol only to recover to normal levels after several hours. These observations suggest that some quite powerful modulation of the HABR is possible, but the mechanism remains completely unknown (see discussion on hydrogen sulfide below).

Although we have shown that adenosine causes postjunctional inhibition of nerve-, norepinephrine-, angiotensin-, and vasopressin-induced constriction of the hepatic artery (but not portal vein), this effect likely occurs only at pharmacological doses of adenosine, with no evidence that such inhibition occurs under physiological conditions [217].

Another puzzling aspect of this work is that the competitive adenosine receptor antagonist, 8-phenyltheophylline, results in dose-related suppression of autoregulation, the HABR, and the response to exogenous adenosine but does not alter the basal tone of the hepatic artery until the dose of antagonist reaches a concentration that produces full blockade, whereupon higher doses result in massive and prolonged constriction [216]. A similar effect is seen in the superior mesenteric artery [197]. Because our hypothesis involves hepatic arterial tone being affected by increases or decreases in adenosine concentration, the absence of effect of small doses of adenosine receptor antagonists on basal tone is without explanation.

5.2 EXTRINSIC INFLUENCES

Although the HABR is the primary regulator of hepatic arterial blood flow, HA flow is also affected by a number of extrinsic factors, drugs, hormones, and nerves.

5.2.1 Caffeine

Considering that caffeine is the drug with largest worldwide consumption and that it is reported to block adenosine receptors, we tested the possibility that caffeine was capable of modifying the buffer response [199]. We found that at doses compatible with extremely heavy consumption, caffeine was without effect on hepatic or superior mesenteric flow or vascular conductance. Caffeine did not antagonize the buffer response even at doses that produced cardiac arrhythmias, whereas it did produce a noncompetitive antagonism of the vasodilation induced by exogenous adenosine. Interestingly, the maximum effect of caffeine was to suppress 60% of the vasodilation induced by adenosine with a remaining 40% being unsuppressable at any dose. This suggested that adenosine caused dilation by two different mechanisms. One, perhaps cyclic AMP dependent, was not affected by caffeine. Caffeine was without effect on isoproterenol-induced vasodilation. The portion of dilation blocked by caffeine may be related to a calcium channel or cyclic GMP effect, but there are no data to directly support this speculation. We concluded from this study that caffeine was unlikely to affect endogenous hepatic or splanchnic blood flow or intrinsic regulatory parameters in response to acute exposure. Effects of chronic exposure have not been evaluated. Adenosine produces dilation through action on A_2 receptors; caffeine is quite selective for A_1 receptors (see Chapter 13).

5.2.2 Vasodilators

Isoproterenol (β_2 receptor), adenosine (A2 receptor), and glucagon are all dilators of the hepatic artery when the drugs are administered directly to the artery. However, if administration is intravenous, direct actions on the splanchnic arteries will lead to increased portal blood flow that, in turn, will activate the buffer response and can actually result in a decrease in flow, depending on the dose [208]. Blocking adenosine receptors in the liver leads to blockade of autoregulation, the buffer response, and the dilator response to adenosine but does not alter the vasodilator responses to other compounds such as isoproterenol.

The hepatic arterial resistance vessels constrict in response to activation of hepatic sympathetic nerves, administration of norepinephrine, vasopressin, and angiotensin. Adenosine is not only a direct-acting vasodilator but it also results in dose-related ability to completely inhibit the vasoconstrictions induced by these stimuli (Figure 5.4).

Adenosine modulation of the nerve-induced constriction of the hepatic artery is likely to be postsynaptic because the constrictor response in the hepatic artery is eliminated by adenosine, but the portal constriction is not simultaneously significantly affected [217]. The importance of the ability of adenosine to interfere with general vasoconstrictors is seen by the protective effect that the buffer response has during hemorrhage where portal blood flow is reduced substantially with

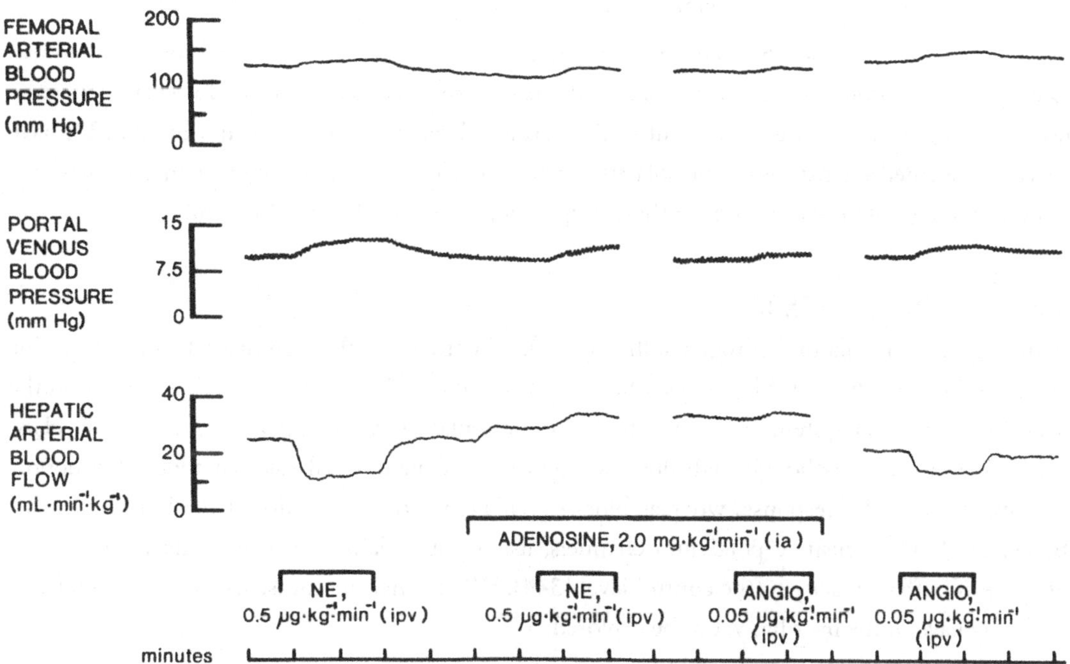

FIGURE 5.4: Trace from one animal showing complete abolition of norepinephrine- and angiotensin-induced vasoconstriction to a high intra-arterial (ia) dose of adenosine. The small rise in hepatic arterial blood flow produced by the constrictions during adenosine infusion is entirely accounted for by the small rise in systemic arterial pressure. Note that no effects on basal portal pressure (portal flow was constant at all measured points) or on the portal constriction induced by nerves, norepinephrine (NE), and angiotensin (ANGIO) were seen at any dose of adenosine tested (blocked ANOVA). Reproduced with permission from Lautt WW, Legare DJ. Adenosine modulation of hepatic arterial but not portal venous constriction induced by sympathetic nerves, norepinephrine, angiotensin, and vasopressin in the cat. *Can J Physiol Pharmacol* 64(4): pp. 449–454, 1986. © 2008 NRC Canada or its licensors. (This figure from publication Can J Physiol Pharmacol is reproduced with permission from publisher NRC Canada).

the resultant accumulation of adenosine serving to dilate the hepatic artery and, at least partially, counteract the effects of blood-borne and nerve-induced vasoconstriction [227].

5.2.3 Carbon Monoxide

Superoxide anions, H_2O_2, nitric oxide (Chapter 9), hydrogen sulfide, and carbon monoxide (CO) are active oxygen species that are produced endogenously and exert their biological actions in the liver. CO, a product of heme oxygenase, upregulates cyclic GMP via activation of guanylyl cyclase and thereby shares several biological actions with NO. Although there has not yet been a proposed specific role for CO in the regulation of the hepatic circulation, suppression of endogenous CO generation results in an increase in the vascular resistance of the liver [354]. Although it has been suggested that the endogeous level of CO generation is the major determinant of steady-state vascular resistance in the liver, the evidence is rather sparse for such a claim. Vascular effects of CO appear to be primarily through effects at the sinusoidal level acting on stellate cells. However, it must be noted that these studies are carried out in the perfused liver, and Greenway and Stark [122] have previously warned against the use of isolated or arterial perfused vascular circuits to imply functional roles in vivo because the responses in these preparations are considerably different.

5.2.4 Hydrogen Sulfide

A third gaseous mediator, hydrogen sulfide (H_2S), is synthesized through degradation of cysteine by cystathionine-γ-lyase (CSE) or cystathionine-β synthase (CBS), both of which are found in the liver. In the vascular system, including the hepatic vasculature, CSE appears to be the only H_2S-generating enzyme. Although H_2S does not appear to regulate basal vascular tone, H_2S donors potentiate the buffer response, whereas blockade of H_2S formation reduced the buffer capacity. Blockade of ATP-sensitive potassium channels, using glibenclamide, reversed the H_2S-induced increase of buffer capacity to the control level [344]. What other neurovascular modulating effects H_2S may have in the liver has yet to be explored.

• • • •

CHAPTER 6

Resistance in the Venous System

Twenty-five percent of the cardiac output flows through the liver with at least two thirds of this derived from the portal vein. This huge portal venous flow is driven through the liver across a minute pressure gradient. The pressure gradient between the portal inflow to the liver and the hepatic venous outflow from the liver is usually no more than 5 mmHg. Precise measurements must take into account the level of the tip of the recording catheter as the small gravitational differences can be significant. The resistance to blood flow through the portal vein is so low because of the unique hepatic vasculature, with conducting blood vessels terminating in each of the microvascular units of the acinus and flowing past only approximately 20 hepatocytes before exiting into the wide hepatic venules. The resistance is so low that at least 50% of the entire blood content of the liver can be expelled without adding significant vascular resistance.

For years, it was believed that the major site of resistance to portal blood flow was at the portal venous inlet vessels. By this assumption, sinusoidal pressure was better represented by hepatic venous outflow (or inferior vena caval) pressure than by portal pressure. However, in 1981, Greenway discussed the new data that changed his mind to the view that the primary site of portal vascular resistance at rest was across the hepatic veins and that portal venous pressure was a more accurate index of mean sinusoidal pressure. He demonstrated zones of resistance identifiable in the small hepatic veins through the use of catheters with the tips sealed and measuring blood pressure through side holes. We later used the same techniques and further validated Greenway's conclusion. In addition, we demonstrated the extreme distensibility of these resistance sites and described the mathematical relationship between the distending pressure of the venous blood and the vascular resistance at the distensible resistance sites in both the hepatic vein and portal venule inlets.

A small change in vascular distending pressure resulted in a large change in the resistance that was defined by the equation $R = 1/P_d^3$, where R is the resistance, calculated based on the pressure on either side of the distending tissue, divided by blood flow across the resistance site. Distending pressure (P_d) is calculated as the mean of the pressure above and below the resistance site. By this equation, a linear relationship is plotted between resistance and $1/P_d^3$ with the intercept passing through zero. The slope of the line is the index of contractility and is calculated as $IC = R \times P_d^3$. My 13-year-old daughter, Kelly, solved the equation (Figure 6.1).

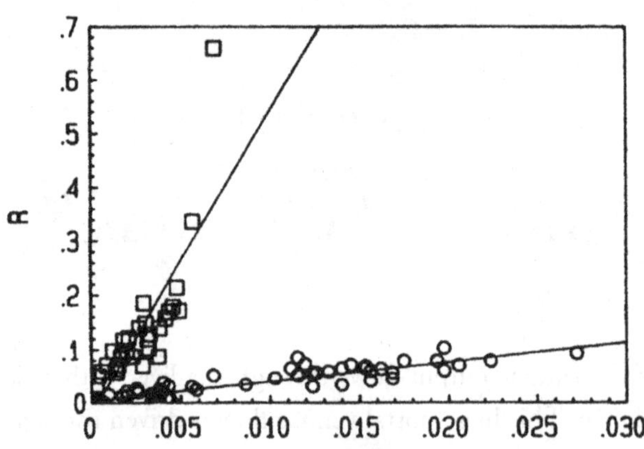

FIGURE 6.1: Demonstration in one cat of the passive relation between total vascular resistance (isolated in situ liver perfused only through portal vein) and distending blood pressure calculated as mean of portal venous and hepatic outflow venous pressure. Data were obtained by changing blood flow over the range of 10–50 ml/min/kg in several steps at normal (3 mmHg) venous pressure and raised (6 and 9 mmHg) venous pressure and by holding flow steady and raising outflow pressure in steps from 3 to 15 mmHg in the control state and during norepinephrine infusion (1.25 μg/min/kg). Because the IC curves extrapolate to zero, IC can be calculated from one data point. Norepinephrine resulted in a large increase in IC (slope), but the passive relationship remained unaltered and was linearly related (IC = $R \times P_d^3$). Reproduced from Lautt WW, Greenway CV, Legare DJ. Index of contractility: quantitative analysis of hepatic venous distensibility. *Am J Physiol* 260: pp. G325–G332, 1991. (This figure from publication Am J Physiol is reproduced with permission from publisher).

The concept of IC allows for determination of whether changes in portal vascular resistance (or any venous resistance for that matter) are active, or passive secondary to altered distending pressures. The passive nature of the distensibility explains what we have referred to as portal pressure autoregulation. Autoregulation generally refers to regulation of blood flow, whereas, in this case, the reference is to regulating portal pressure. The liver cannot control portal blood flow, which is simply the total venous efflux from all of the splanchnic organs. Hepatic venous resistance is so low that even maximal stimulation of the hepatic sympathetic nerves, although resulting in a doubling or tripling of portal venous pressure, does not alter portal outflow from the splanchnic organs and therefore does not alter portal inflow to the liver. The liver must accommodate the entire portal flow and must accommodate to the quite large fluctuations that occur in portal blood flow, while maintaining its other homeostatic functions. The passively distensible portal venous and hepatic venous resistance sites explain how it is possible for portal blood flow to double while producing no more than a 2-mmHg change in portal pressure.

The resistance to blood flow through the liver is extremely low with pressure gradients between the portal venous inflow and hepatic venous outflow of the liver being in the range of 5 mmHg or less. Considering that the pressure gradients across all other organs are in the range of 115 mmHg, the vascular resistance within the hepatic portal system might erroneously appear to be trivial and of no consequence. In fact, the pressures are regulated at precise regions of the venous circuit at presinusoidal and postsinusoidal sites, and the proportion of pressure drop across the two sites and the total resistance determines intrahepatic and portal venous pressure and volumes. In chronic diseases of the liver, the most common cause of death is directly related to the vascular consequences of increased resistance to portal blood flow. If the pressure gradient becomes elevated by as little as 15 mmHg, hemodynamic and homeostatic instability can lead to the demise of the patient. Portal venous pressure reaches approximately 25 mmHg in the most severely cirrhotic conditions [368].

The adjustment of pressure in the normal state is achieved by changes of vascular resistance within the presinusoidal portal venules and the postsinusoidal hepatic veins. These sites of resistance have three important characteristics: they offer very low resistance; the resistance is able to be more than doubled by active vascular constriction; the resistance sites are passively distensible. To understand the venous responses to vasoactive stimuli, it must first be appreciated that the active responses interact with and are severely modulated by the passive distensibility of the resistance site. The ability to describe this interaction is dependent on being able to measure resistance at the key regulatory locations.

6.1 ESSENTIAL ASSUMPTIONS

The description of active and passive regulation of presinusoidal and postsinusoidal resistance sites is dependent on the assumption that the pressures measured are valid and not subject to significant artifact. The pressures of relevance are the portal venous pressure before entry into the liver, the central venous pressure (CVP) at the exit of the hepatic veins, and the intrahepatic pressure representing sinusoidal blood pressure. With these three pressures, and portal and hepatic blood flows, the presinusoidal and postsinusoidal resistance can be quantified. The portal and vena caval pressures are technically easily measured and validated because of ready access to these vessels under experimental conditions. The intrahepatic pressure is the contentious pressure.

As a pressure catheter is advanced via the vena cava into the hepatic veins, pressure is similar to CVP until the catheter tip passes through a narrow length of hepatic vein (Figure 6.2). This region has the characteristics of a smooth muscle sphincter, and morphological studies support the existence of sphincter-like regions in large hepatic veins [52]. These sphincters are localized to third-order branches (ramuli, according to the nomenclature of Elias and Petty [75]) of the hepatic veins in cats [215] and in the terminal 2 cm of the lobar hepatic vein proper in dogs [236]

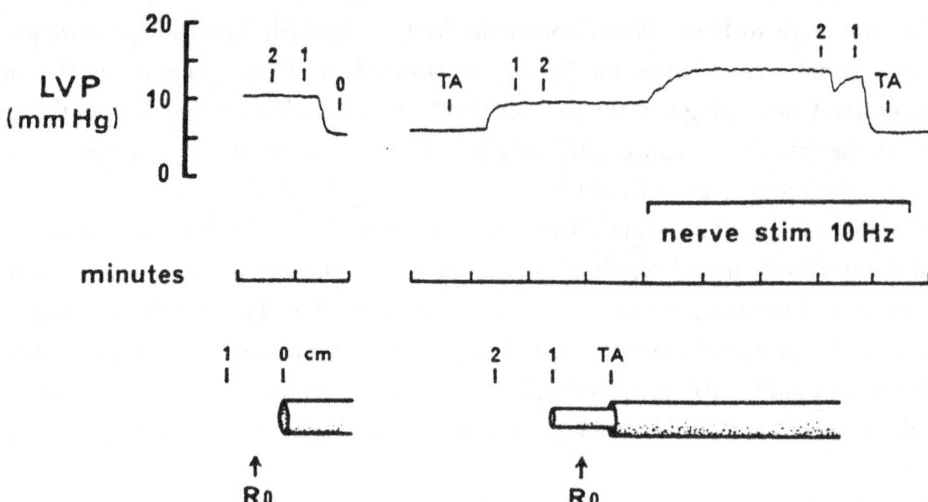

FIGURE 6.2: A hepatic venous pressure profile was obtained by withdrawing a large (PE240, external diameter 2.42 mm) catheter from a wedged position in the hepatic veins in 1-cm steps (left diagram) until recorded pressure (LVP-labor venous pressure) dropped to near CVP. The location of the resistance site (sphincter) is indicated by the large pressure drop (RO). With the large catheter left in place (within 1 cm of RO) and tied at the jugular vein, a small catheter (PE90, external diameter 1.27 mm) was passed down the inside of the large catheter until the tips aligned (TA). Pressure was monitored via the small catheter, which was then advanced into the hepatic vein and seen to pass the RO site within the first centimeter. At 2 cm proximal to RO, the LVP was similar to what it had been when recorded using the large catheter. The hepatic nerves were then stimulated at 10 Hz and the LVP rose in response. As the catheter was withdrawn in 1-cm steps, the major site of resistance (RO) was still seen across the vascular segment that provided the resistance in the control state. This test demonstrates that the pressures measured as LVP are not significantly dependent on the size of the catheter and that the RO site can be consistently located relative to a constant point (the fixed catheter). Similar pressures and profiles have been measured using small catheters with sealed tips recording from side holes. Reproduced with permission from Legare DJ, Lautt WW. Hepatic venous resistance site in the dog: localization and validation of intrahepatic pressure measurements. *Can J Physiol Pharmacol* 65(3): pp. 352–359, 1987. © 2008 NRC Canada or its licensors. (This figure from publication Can J Physiol Pharmacol is reproduced with permission from publisher NRC Canada).

(Figure 6.3). Physiological sphincters have not been proven in other species, and morphological data directly correlating functional and structural evidence are missing. The sudden rise of measured pressure from CVP to portal pressure as a catheter is advanced has usually been assumed to represent wedging of the catheter, and the pressure readings were assumed to be via a static column. Several factors indicate that this measurement is not a "wedged" pressure. First, the catheter need not be wedged and can often be advanced at least an additional 1–2 cm beyond the sphincter even in cats [215]. Second, the catheters we have used have the tips sealed and pressure is recorded via side holes cut 3 and 7 mm back from the tip. We assume that the ability to record a valid pressure beyond the sphincter, but distal to the sealed catheter tip, is dependent on the existence of collaterals between the veins proximal to the sphincters. The sphincters are contracted by neural and pharmacological stimuli, and the site of the sphincter is not altered by catheter size or state of contraction [215,223,236] as would be expected if the "sphincter" really represented an artifact of wedging in the vein (Figures 6.2 and 6.4). Finally, in dogs, an unusual species in that the hepatic resistance site is located in the terminal portion of the hepatic vein, it is possible to position a catheter in the hepatic vein proximal to the sphincter site via an incision in the surface of the liver. By locating a

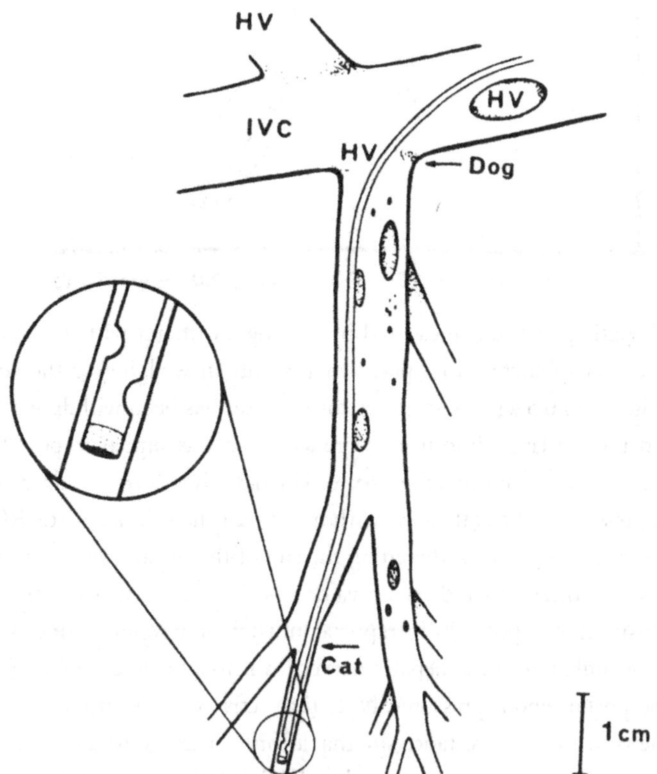

FIGURE 6.3: In cats, the resistance site is localized to ramuli that drain into venous rami and then to the trunk of hepatic vein (HV). Pressure distal to these ramuli is insignificantly different from pressure in the inferior vena cava (IVC). In dogs, pressure in the large trunk is similar to portal venous pressure and the resistance site is located within 1–3 cm of the junction of IVC and HV. Reproduced from Lautt WW, Legare DJ. Effect of histamine, norepinephrine and nerves on vascular resistance and pressures in dog liver. *Am J Physiol* 252: pp. G472–478, 1987. (This figure from publication Am J Physiol is reproduced with permission from publisher).

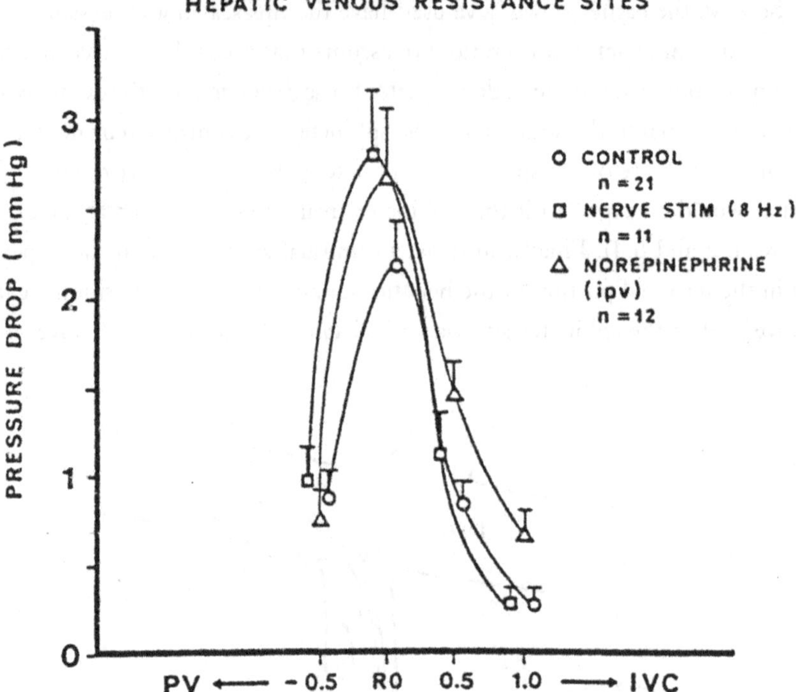

FIGURE 6.4: Intrahepatic pressure is measured by passing a catheter with a sealed tip and side holes beyond the hepatic venous sphincters to a point that results in wedging of the venous catheter. The catheter is then withdrawn in 0.5-cm steps. After the catheter has been withdrawn from 1 to 2 cm, the pressure recorded undergoes a large drop from a pressure that was similar to portal venous (PV) pressure down to a pressure similar to vena caval pressure. The data shown are standardized to plot the mean decrease in measured pressure as the catheter is withdrawn past the sphincter site. RO is standardized as the area of greatest pressure drop and is, therefore, the site of the hepatic venous sphincter. Most of the pressure drop between the portal vein and vena cava occurs across a vascular segment less than 0.5 cm in length. Active vasoconstriction imposed by intraportal infusion of norepinephrine (1.25 μg/kg/min), and the response to 8-Hz stimulation of the hepatic nerves leads to a hepatic venous sphincter constriction and a rise in lobar and portal venous pressures. Note that active vasoconstriction caused hepatic venous constriction across the same narrow resistance site that afforded basal resistance. The points on the graph are slightly offset so that the symbols do not overlap. IVC, inferior vena cava. Reproduced from Lautt WW, Greenway CV, Legare DJ. Effect of hepatic nerves, norepinephrine, angiotensin, elevated CVP on postsinusoidal resistance sites and intrahepatic pressures in cats. *Microvascular Research*, vol. 33, no. 1, p. 55. © 1987 by Academic Press. (This figure from publication Microvascular Research is reproduced with permission of publisher Academic Press).

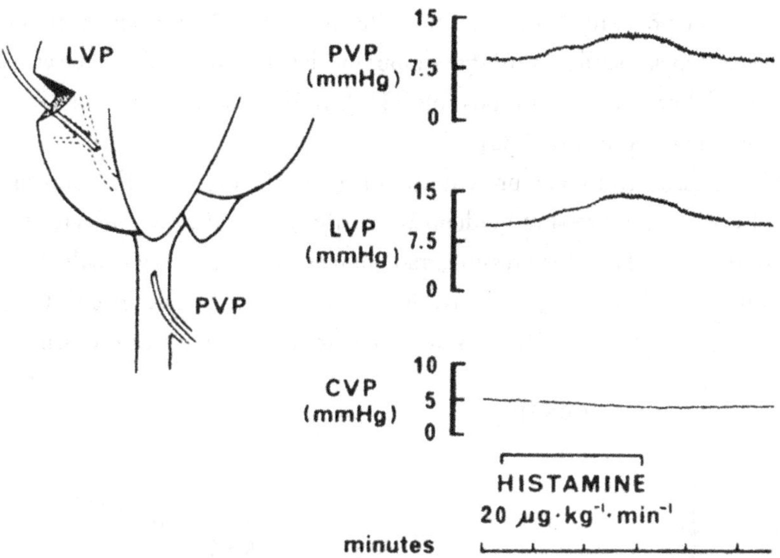

FIGURE 6.5: Trace from one dog showing labor venous pressure (LVP) measured by passing the catheter through a slash in the liver lobe, via a small hepatic vein (1–2 mm diameter) into the large hepatic vein (3–5 mm diameter). Pressure measured throughout the vessel did not change and was similar to simultaneously measured portal venous pressure (PVP). The response to intraportal infusion of histamine showed LVP and PVP to change in parallel and confirmed the results obtained using the LVP catheter passed into the hepatic veins through the hepatic venous resistance site via the vena cava. This test shows clearly that the entire portal pressure response to histamine can be attributed to constriction of the hepatic venous sphincter downstream from the catheter tip located in the large hepatic vein. Using this technique it is not possible that the LVP measurement was reflective of a wedged pressure or other artifact because the LVP catheter remained upstream of the venous sphincters. Reproduced with permission from Legare DJ, WW Lautt. Hepatic venous resistance site in the dog: localization and validation of intrahepatic pressure measurements. *Can J Physiol Pharmacol* 65(3): pp. 352–359, 1987. © 2008 NRC Canada or its licensors. (This figure from publication Can J Physiol Pharmacol is reproduced with permission from publisher NRC Canada).

small hepatic venous tributary and passing a catheter downstream into the hepatic vein (the catheter thus does not pass through the putative sphincter zone), pressure similar to portal pressure can be demonstrated (Figure 6.5). With the catheter thus placed, histamine-induced sphincter contraction led to equal and parallel elevations in portal venous pressure and in hepatic venous pressure proximal to the hepatic veins in a position where it is impossible to be measuring a "wedged" pressure [236]. Other arguments have been presented [114,213,218] to support the view that measurement of lobar venous pressure (LVP) using a catheter that is passed proximal to the hepatic venous resistance site is representative of a true pressure measured at that site. Use of a balloon occluder in the large hepatic veins does, however, block a sufficiently large venous outflow that pressure escape via interconnected venous and sinusoidal pathways is not possible [224]. In that case, the balloon occluder measures pressure resembling portal pressure [222].

Thus, LVP is assumed to be representative of sinusoidal pressure. The pressure gradient between LVP and CVP is the pressure gradient primarily regulated by a sphincter-like zone in the hepatic veins. In the basal state, this pressure gradient usually represents virtually the entire pressure gradient across the liver. Therefore, in the basal state, the vascular resistance of the portal venules and sinusoids is normally trivial with sinusoidal pressure and portal pressure being insignificantly

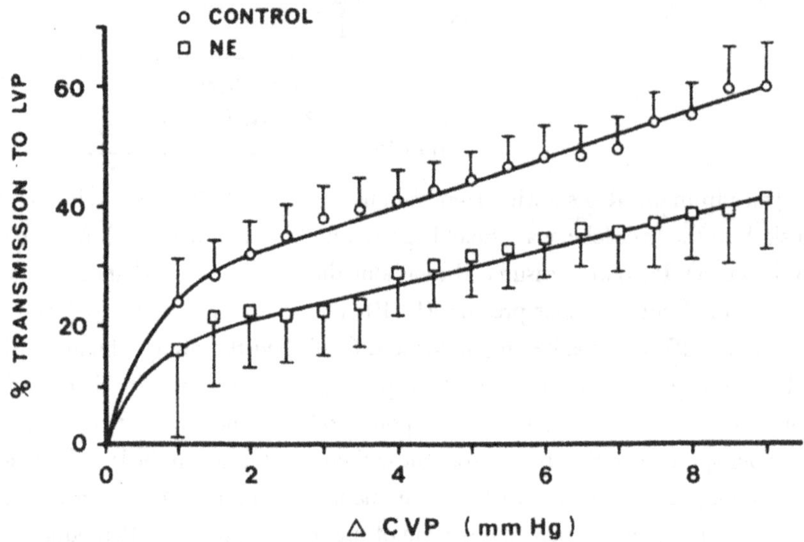

FIGURE 6.6: The percentage of a rise in CVP transmitted past the hepatic sphincter to the upstream LVP site. The control state led to transmission of the rise in CVP to LVP according to the following equation derived from linear regression of all data: percent transmission to LVP = 23.4 + 4.0 CVP; $r = .99$. The percent transmission increased at greater distending pressures until at an increase in CVP of 9 mmHg, 59% transmission occurred. In the presence of hepatic venous sphincter constriction imposed by norepinephrine, transmission of raised CVP was less (percent transmission to LVP = 15.3 + 2.9 CVP; $r = .98$; equation derived from linear regression of all data). At a rise in CVP of 9 mmHg, the percent transmission rose to approximately 41%. The slope of the lines was significantly different ($P < .001$). There is no evidence of any critical pressure below which no transmission occurs in contrast with the response predicted by the waterfall model. Reproduced with permission from Lautt WW, Legare DJ, Greenway CV. Effect of hepatic venous sphincter contraction on transmission of CVP to lobar and portal pressure. *Can J Physiol Pharmacol* 65(11): pp. 2235–2243, 1987. © 2008 NRC Canada or its licensors. (This figure from publication Can J Physiol Pharmacol is reproduced with permission from publisher NRC Canada).

different. The observation that equal reductions in arterial and portal flow lead to the same changes in hepatic volume [19] supports the contention that virtually all of the resistance to venous flow is in the hepatic veins. The pressure gradient between the portal vein (PVP) and the pressure (LVP) measured just proximal to the hepatic venous sphincters is assumed to represent primarily resistance within the portal venules. Note, however, that the pressure gradient represents pressure lost across the large and small portal tributaries, and the sinusoidal bed, as well as the small hepatic venules proximal to the LVP catheter. Active vasoconstriction results in a very significant increase in the PVP to LVP gradient.

I will refer to this pressure gradient as being due to presinusoidal or portal venous resistance. Using these assumptions to analyze the vascular responses to norepinephrine and nerve stimulation reveals some interesting characteristics. For example, in cats, activation of sympathetic nerves results in an initial vasoconstriction of the hepatic arteriole and portal venule sites, which gradually undergo vascular escape, in contrast to the responses of the hepatic venous resistance sites that actively contract and maintain contracted throughout the period of nerve stimulation. Vascular escape of the presinusoidal portal resistance vessel sites does not, however, occur in response to norepinephrine infusion in contrast to the escape seen in the hepatic artery [218]. The observation of almost complete vascular escape of the active nerve-induced constriction in the portal vein indicates that a very small pressure gradient between the portal vein and hepatic vein existed after 5 min of stimulation. The reduction in hepatic blood volume, which is well maintained, obviously did not significantly elevate vascular resistance. This observation suggests that the sinusoidal cross-sectional area is so vast as to offer trivial resistance and that very large changes in this cross-sectional area can be achieved without adding significantly to overall resistance.

Changes in CVP are transmitted past the venous sphincter to the LVP according to the distensibility of the resistance site. Increasing the IC using norepinephrine results in a reduced transmission (Figure 6.6).

In summary, the principle assumptions here are that the LVP measurement is a valid pressure measurement representative of pressures proximal to the hepatic venous sphincter-like zone and that the PVP–LVP gradient represents largely presinusoidal or portal venous resistance and the LVP–CVP gradient represents primarily resistance across the hepatic venous sphincter-like zones.

6.2 PASSIVE DISTENSIBILITY

The large passive distensibility of the venous resistance sites has been quantified and studied under a variety of conditions [214,218,219]. Hepatic venous resistance is reduced by 75% in response to an elevation of vena caval pressure of roughly 3 mmHg [231]. The small rise in distending pressure produces very large decreases in vascular resistance that has been quantified as having a constant relationship under basal condition and under conditions of actively increased basal tone. The distending pressure (P_d) is able to be estimated from the average value of the pressures measured on either side of the resistance site [113]. Resistance (R) is inversely linearly related to the distending pressure cubed (P_d^3) and the slope of this linear relationship is referred to as the "index of contractility" where IC = $R \times P_d^3$. Changes in this IC are affected acutely only by active vascular responses and are independent of passive alterations in the distending blood pressure. Thus, active vascular responses can be measured using the IC, whereas the use of calculated vascular resistance provides a measure that is the result of interaction between active and passive influences (see Chapter 11, Hepatic Nerves).

The utility of the IC can be seen when attempting to interpret the vascular responses to pharmaceutical compounds given to modify portal vascular resistance, for example, for treatment of portal hypertension. A compound such as propranolol that reduces portal pressure secondary to reductions of cardiac output and portal inflow actually result in an increase in calculated vascular resistance, thus giving the false impression that propranolol results in active vasoconstriction of the portal vessels. The calculated increase in resistance is, however, most likely explained entirely by the reduction in portal flow and subsequent portal pressure and passive recoil of the resistance sites. Using the IC, an increase in resistance with no change in IC offers a clear differentiation between active and passive responses. Similarly, pharmaceutical evaluation of compounds capable of dilating resistance sites in portacaval shunts, as a means of reducing portal pressure, is best assessed using both calculated resistance and IC.

• • • •

CHAPTER 7

Fetal and Neonatal Hepatic Circulation

The fetal circulation and its transition to the neonatal circulation is an elegant symphony of homeostatic coordination. The fetal circulation has several features that are lost at birth. The ductus arteriosus, ductus venosus, the foramen ovale, and the placenta are all essential for fetal survival in an environment where the fetus is entirely dependent on the maternal circulation to provide for every need from the outside world. Whereas the lungs receive the entire cardiac output after birth, the collapsed lungs in the fetus receive approximately 10% of the cardiac output. The placenta is the "fetal lung." The placental villi reach into the rich uterine vascular bed and contain small terminal branches of the fetal umbilical arteries and vein and capillary exchange vessels. Fifty-five percent of the fetal cardiac output goes through the umbilical artery to the placenta. Blood from the placenta passes back to the fetus through the umbilical vein, which has an oxygen saturation of approximately 80% compared with the 98% saturation in the arterial circulation of the adult. The ductus venosus (Figure 7.1) carries some of this blood, bypassing the liver, directly to the inferior vena cava where it joins blood from the lower trunk and extremities (26% oxygen saturated) and from the liver. The blood from the liver and the ductus venosus enter the inferior vena cava and pass to the right atrium. Most of the blood entering the heart through the inferior vena cava is diverted directly to the left atrium via a patent foramen ovale in the atrial septum. Most of the blood from the superior vena cava enters the right ventricle and is expelled into the pulmonary artery. The resistance of the collapsed lungs is high, and the pressure in the pulmonary artery is higher than it is in the aorta, so that most of the blood in the pulmonary artery passes through the ductus arteriosus to the aorta. The less saturated blood from the right ventricle is thus diverted to the trunk and lower body of the fetus, whereas the head of the fetus receives the better oxygenated blood from the left ventricle. From the aorta, approximately 60% of the fetal cardiac output is pumped into the umbilical arteries and back to the placenta.

The low oxygen saturation of the fetal circulation is a significant and important characteristic. Fetal red blood cells contain fetal hemoglobin, which has a substantially higher affinity for oxygen than does the hemoglobin in adult red blood cells. Before birth, the left and right sides of the heart pump in parallel in the fetus rather than in series as they do in the adult. At birth, the placental circulation is eliminated and pressure in the aorta rises. Because the placental circulation is absent,

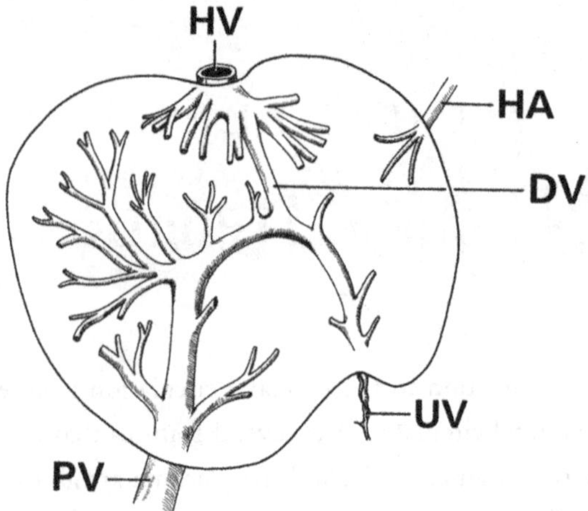

FIGURE 7.1: Most of the oxygenated blood reaching the heart via the umbilical vein and inferior vena cava is diverted through the foramen ovale and pumped out the aorta to the head, whereas the deoxygenated blood, returned via the superior vena cava, is mostly pumped through the pulmonary artery and ductus arteriosus to the feet and the umbilical arteries. Reproduced from Ganong WF. *Review of Medical Physiology*. Appleton & Lange, East Norwalk, Connecticut. Chapter 32, p. 581, 1991. (This figure from publication Review of Medical Physiology is reproduced with permission from publisher Appleton & Lange/McGrawHill).

the infant becomes increasingly hypoxic and finally gasps, causing the lungs to expand. The negative interplural pressure and the constriction of the umbilical veins squeeze as much as 100 ml of blood from the placenta, which is referred to as the placental transfusion. Once the lungs are expanded, the pulmonary vascular resistance falls to less than 20% of the in utero value and pulmonary blood flow increases markedly. Blood returning from the lungs raises the pressure in the left atrium, thus closing the foramen ovale by pushing the valve against the interatrial septum. In response to the increased oxygen tension in the blood flowing through the ductus arteriosus, the ductus arteriosus constricts. The constriction is initiated within a few minutes after birth but may not close completely for a few days. The foramen ovale and the ductus arteriosus both fuse shut in normal infants and by the end of a few days of life the adult circulatory pattern is established. At birth, the umbilical vein constricts and eventually appears as a small remnant component of the falciform ligament that connects the liver to the diaphragm and peritoneal cavity lining.

Joe Zink, who trained with Clive Greenway, carried out an elegant series of experiments in lamb fetuses and neonates with the intention of resolving decades-old controversies related to the unique hepatic circulation in the perinatal stage. He used microsphere distribution protocols to evaluate regional blood flows. The fetal liver shows functional heterogeneity of the parenchymal cells dependent on the microcirculation. The right side of the liver is perfused with less well-oxygenated blood, mainly from the portal blood, and has a greater role in hematopoiesis. The left side of the liver receives blood mainly from the umbilical vein, has a higher content of oxygen-dependent enzymes, and is more active in drug binding and metabolism. In fetal sheep, 80% of the total hepatic blood flow was from the umbilical vein with the remainder almost exclusively supplied by the portal blood flow, with trivial contribution from the hepatic artery. In 1-day-old lambs, the hepatic arterial flow increased 18-fold compared to the flow in lamb fetuses. Portal blood flow accounted for approximately 86% of total hepatic blood flow; however, a significant fraction (20% of the portal blood flow) was diverted away from the liver circulation via the ductus venosus. After birth, the portal flow became uniformly distributed throughout the organ. In lambs, the ductus venosus remained patent for 2–3 days after birth but the flow was highly variable. The pattern of ductus venosus closure reported by Zink [384] has also been noted in humans [241,245].

• • • •

CHAPTER 8

In Vivo Pharmacodynamic Approaches

To study any phenomenon, it is useful to have a quantifiable index. The index must have a direct relevance to the studied phenomenon and the measured index should be expressed in a useful manner. The vascular resistance of blood vessels has been the subject of speculation and investigation at least since Laplace described a mathematical expression for vascular resistance. Vascular resistance is defined as the pressure gradient across the vascular bed, divided by the flow through the bed. I will discuss the merits of quantifying arterial vascular tone using resistance or, more appropriately, conductance (flow divided by pressure gradient).

The vascular tone in blood vessels is affected by both active and passive forces. Constriction of vascular smooth muscle results in an active increase in vascular resistance to blood flow. An increase in blood pressure will result in a passive elastic stretch of the resistance vessels. In this chapter, I also discuss the use of an index of contractility that finds major utility in describing the active and passive elastic forces that influence venous resistance sites. These concepts have been developed for the liver, but use of the index of contractility also found utility in elegantly describing the cardiovascular system of the octopus [2]. The differentiation between venous resistance and the index of contractility has also been applied to portacaval shunts that form in response to chronic elevation in intrahepatic venous resistance [143]. With the use of these two indices of vascular tone, I will then describe general procedural approaches for establishing the experimental setup to permit in vivo pharmacodynamic studies to be carried out.

8.1 RESISTANCE OR CONDUCTANCE?

Traditionally, vascular tone has been expressed in units of vascular resistance. Stark [351] and Robard [317] argued for the use of vascular conductance, the inverse of resistance, to quantitate vascular tone. In my first studies on the hepatic artery, I adopted the use of vascular resistance [230], ignoring the wise council of my colleague Ron Stark. But I thereafter used conductance to represent arterial tone [184]. At that point, there did not appear to be an obvious advantage to using conductance over resistance except for the theoretical points raised by Stark.

Pharmacodynamic relationships cannot be described from arterial resistance responses, whereas vascular responses, expressed as conductance, are readily quantitated. The most obvious example is pharmacodynamic studies related to a potent vasoconstrictor. As the intensity of the constriction increases, blood flow decreases and may actually cease. As blood flow approaches zero, calculated resistance approaches infinity, a quantitative value that is useless. Arterial conductance, however, falls along with blood flow, and at zero flow, conductance is zero. At constant perfusion pressure, conductance is linearly related to blood flow. Local changes in vascular tone produce primarily changes in blood flow rather than systemic pressure. Both the pressure and flow parameters must be considered when expressing vascular tone because both parameters change simultaneously. However, changes in regional arterial vascular tone largely result in changes in blood flow; therefore, blood flow should be in the numerator of the equation in order that the parameter and the index reflecting the parameter change similarly. Using vascular conductance, pharmacodynamic calculations can be done using enzyme kinetic mathematics.

The data analyzed in Figures 8.1 and 8.2 were obtained from the superior mesenteric artery. Electrical stimulation of the periarterial nerve bundle at 1, 3, and 9 Hz was carried out in the presence and absence of two doses of intra-arterial adenosine. Both resistance and conductance curves show clearly that adenosine antagonizes the nerve-induced vasoconstriction. Attempts to quantitate the degree and type of antagonism, however, are highly dependent on the means of data expres-

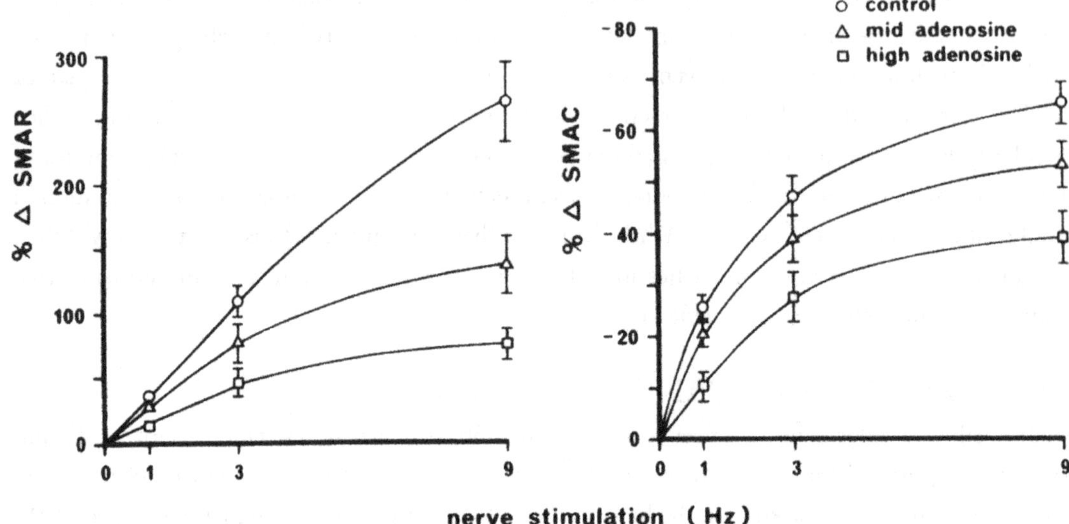

FIGURE 8.1: Frequency–response relationships of sympathetic nerve stimulation in the superior mesenteric artery expressed as percentage change in resistance (% SMAR) or conductance (% SMAC) in the control state and during intra-arterial infusion of two doses of adenosine (0.05, 0.4 mg/kg/min). Reproduced from Lautt WW. Resistance or conductance for expression of arterial vascular tone. *Microvasc Res* 37: pp. 230–236, 1989. (This figure from publication Microvasc Res is reproduced with permission from publisher).

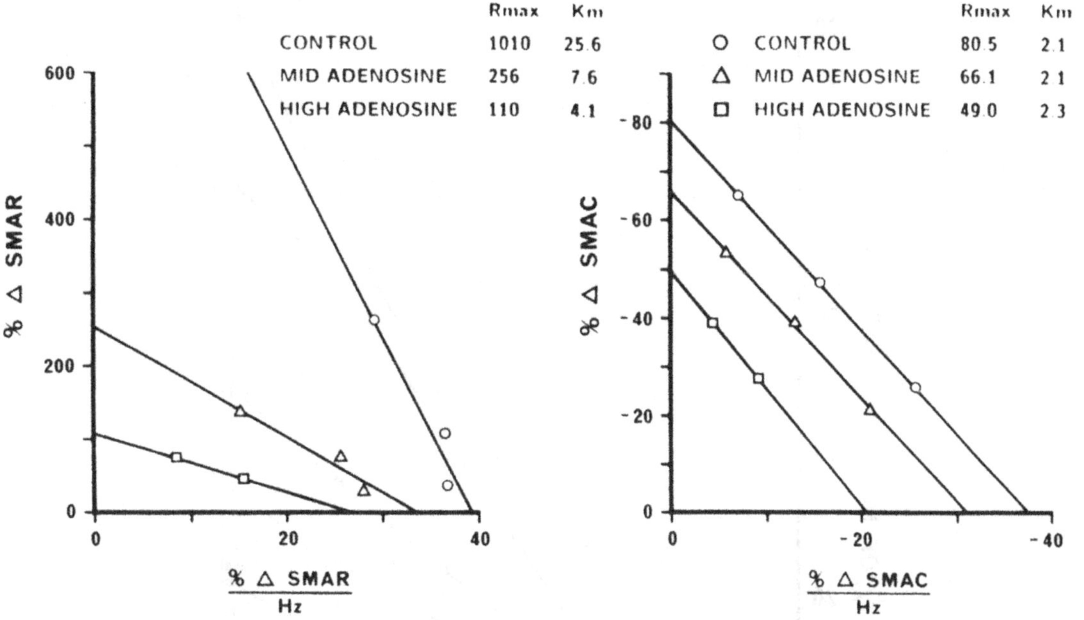

FIGURE 8.2: Data from Figure 8.1 linearized by Eadie–Hofstee plots using vascular responses expressed as percentage change in resistance (% SMAR) or conductance (% SMAC) showing calculated values of the maximal vasoconstriction (R_{max}) and K_m (or more appropriately the Hz_{50}), the frequency of nerve stimulation that will produce half of the R_{max}. K_m is not affected by adenosine, but R_{max} is dose-dependently suppressed indicating classic noncompetitive antagonism. The use of calculated resistance provides no value and greatly distorts the calculated parameters. Reproduced from Lautt WW. Resistance or conductance for expression of arterial vascular tone. *Microvasc Res* 37: pp. 230–236, 1989. (This figure from publication Microvasc Res is reproduced with permission from publisher).

sion. When the data are plotted using a standard Lineweaver–Burke (1/response vs 1/stimulus) or Eadie–Hofstee plot (response vs response/stimulus), kinetic parameters of the responses come clear only if conductance is used. Figure 8.2 shows a classic Eadie–Hofstee plot of response (as percentage change in vascular conductance) against response divided by the frequency of stimulation. The intercept on the ordinate indicates the R_{max}, that is, the maximal mean vasoconstriction attainable, which is seen to be equivalent to reduction of conductance of 80%. The K_m, or frequency of stimulation that procures 50% of R_{max}, is seen as the negative slope of the line or can be calculated from the intercept of the abscissa (R_{max}/K_m). In the presence of adenosine, the curve is shifted in parallel, indicating that adenosine produces noncompetitive antagonism of nerve-induced constriction. That is, R_{max} is reduced by adenosine but K_m is unchanged (competitive antagonism would be indicated by constant R_{max} but changing K_m). A clear dose-related effect is seen, as a higher dose of adenosine shifts the curve further. Similar plots of changes in resistance from the same data produced

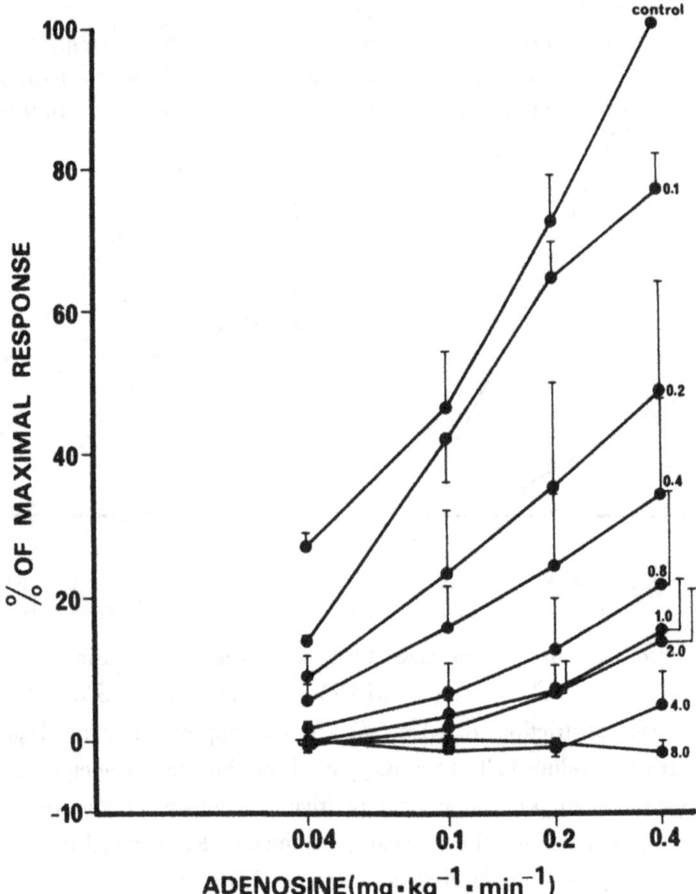

FIGURE 8.3: Dose–response curves for intraportal adenosine infusion in the absence of antagonist (control) and in the presence of stepwise increases in dose (i.a.) of 8-phenyltheophylline (dose is expressed in milligrams per kilogram; by use of cumulative doses of antagonist, an animal receiving a dose of 8 mg/kg will have received all previous doses over about a 3-hour period). These data are reexpressed in Figure 8.4 to estimate pharmacodynamic parameters. Reproduced with permission from Lautt WW, Legare DJ. The use of 8-phenyltheophylline as a competitive antagonist of adenosine and an inhibitor of the intrinsic regulatory mechanism of the hepatic artery. *Can J Physiol Pharmacol* 63(6): pp. 717–722, 1985. © 2008 NRC Canada or its licensors. (This figure from publication Can J Physiol Pharmacol is reproduced with permission from publisher NRC Canada).

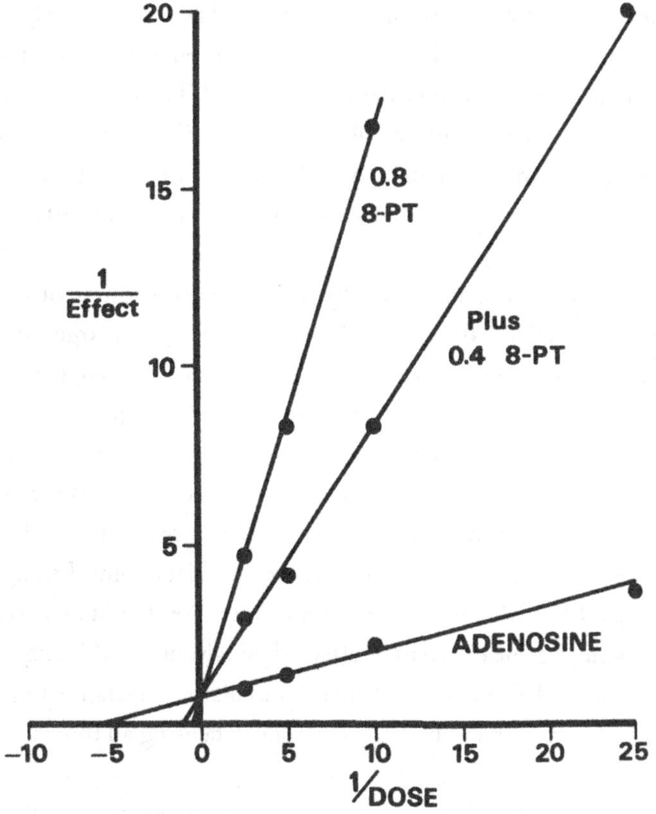

FIGURE 8.4: Double reciprocal Lineweaver–Burke plot from Figure 8.3. Maximal dilation seen is rated as 1, and if a vasodilation of 86% of the maximal response was seen, it would be entered at 1/(0.86). The *y*-intercept estimates maximal vasodilation and, converted to change in arterial conductance from basal (100%), is equivalent to 245%. The dose of adenosine that produces one half of maximal vasodilation is estimated from the *x*-intercept to be 0.19 mg/kg. Progressive increase of dose of 8-phenyltheophylline (8-PT, dose expressed in milligrams per kilogram) shows no shift in maximal dilation but progressive increase in the dose required for one-half maximal dilation, indicating a classic competitive antagonism of adenosine vasodilation (calculated from mean data, $n = 5$). Reproduced with permission from Lautt WW, Legare DJ. The use of 8-phenyltheophylline as a competitive antagonist of adenosine and an inhibitor of the intrinsic regulatory mechanism of the hepatic artery. *Can J Physiol Pharmacol* 63(6): pp. 717–722, 1985. © 2008 NRC Canada or its licensors. (This figure from publication Can J Physiol Pharmacol is reproduced with permission from publisher NRC Canada).

uninterpretable results. The control K_m is calculated as 25.6 Hz, a value clearly not representing the biological reality because maximal vascular responses occur at 6–10 Hz [213]. In addition, both R_{max} and K_m undergo dramatic decreases as the adenosine is added. The kinetics of stimulus antagonism are rendered meaningless by the use of vascular resistance. Similar types of kinetic calculations using the Lineweaver–Burke transformation have been used to show that 8-phenyltheophylline competitively antagonizes adenosine-induced vasodilation in the hepatic artery when vascular tone is assessed using conductance [216].

Although the linear conversion provided by the Lineweaver–Burke or Eadie–Hofstee transformations provides interpretable data, the smallest responses have a large impact on the slope of the relationship. Small errors in the smallest responses can distort the slope. In contrast, computer-assisted iterative solution of the nonlinear rectangular hyperbolic shape of a dose–response curve or nerve response curve allows for calculation of R_{max} and ED_{50} values with as few as three measured data points. The expression of the X-axis on a linear scale rather than a log scale demonstrates that the point of highest confidence in defining the mathematical relationship of these curves is with the zero value. The use of log transferred data loses the zero data point. Ideally, a second response would be approximately 50% of the maximal response and there should be two other stimuli, preferably producing a maximal or near maximal effect. The advantage of being able to use the zero value cannot be overstated. The utility of calculated vascular conductance to provide valid data is demonstrated by the ability to extract pharmacodynamic data using all three methods of estimating R_{max} and ED_{50}.

The ability to carry out studies deriving multiple data points for pharmacodynamic calculations are shown in Figures 8.3 and 8.4. Intraportal adenosine was infused and the response of the hepatic artery was determined to four incremental doses of infusion of adenosine tested against progressively increasing doses of an adenosine receptor antagonist, 8-phenyltheophylline. Figure 8.4 shows the data from Figure 8.3 plotted according to the Lineweaver–Burke double reciprocal plot of the data.

The hepatic capacitance responses can also be analyzed using the nonlinear regression of the rectangular hyperbolic dose–response curve (GraphPad, ISI Software). The Hz_{50} (nerve frequency stimulation required to produce 50% of the maximal response) in the cat liver was 3.4 Hz. Both adenosine and glucagon produced modulation of sympathetic nerve-induced capacitance responses, although neither compound had significant effects on basal blood volume. Adenosine did not affect the Hz_{50} but produced a modest suppression of the R_{max}. In contrast, glucagon produced a modest decrease in R_{max} and an increase in the Hz_{50} [214].

8.2 INDEX OF CONTRACTILITY

The concept of the index of contractility (IC) has been described in Chapter 6. The IC is useful for differentiating changes in venous resistance as active or passive responses. Note that in the case of the hepatic venous resistance sites, changes in venous tone result in changes in venous pressure and not in flow. The pressure should therefore be in the numerator of the equation, and vascular tone in the venous system is best represented using calculations of vascular resistance rather than vascular conductance.

The differentiation between resistance and IC is also a useful tool for hemodynamic studies of vascular control of portacaval shunts that form in portal hypertension. The vascular preparation used for this approach uses a chronic portal venous occlusion model using a vascular constrictor that absorbs fluid and gradually swells shut over a 4-week period. With all portal flow going through the shunts into the vena cava, the pressure gradient acting on these blood vessels can be calculated as the mean of upstream (portal venous) and downstream (IVC) pressures and, if other arterial inputs to the portal vein are ligated (inferior mesenteric artery, gastric, and splenic arteries), the only blood flow is through the superior mesenteric artery, which can be readily quantified [142]. In this preparation the spleen is removed. The areas normally supplied by the occluded arteries have anastomotic connections to the superior mesenteric artery. All flow in the shunts is derived from the superior mesenteric artery.

The IC for portacaval shunts can be calculated in the same manner as done for the hepatic venous resistance sites. Shunt resistance is plotted against 1/distending pressure^3. Because R is linearly related to $1/P_d^{\,3}$ and the intercept passes through zero, the IC can be calculated accurately from a single data point. The advantage that the IC calculation offers is that IC changes only in response to active stimuli, whereas calculated changes in resistance incorporate both active and passive responses. This is especially important in considering the effect of drugs used to treat portal hypertension. The one point determination of IC can be validated by demonstrating that IC does not change passively when blood flow (pressure) is changed passively. This type of portacaval surgical preparation can also be done in rats [279].

In the liver, a large active venoconstriction will result in an increased upstream venous pressure which, in turn, will act as a distending pressure on the compliant venous resistance sites, thereby counteracting the active constriction. The vasoconstrictor response to doses of norepinephrine demonstrates the difference in impression provided by the different indices. In response to an intraportal infusion of 1.25 μg/kg/min of norepinephrine, the IC of the portal resistance vessels rose by 89%, whereas the resistance increased by only 26% because the distending pressure had also increased by 14% [143].

8.3 SURGICAL PREPARATION CONSIDERATIONS

For pharmacodynamic studies, accurate dose–response relationships can be obtained by measurements of blood flow in the artery, the inflow and outflow blood pressures, and access to an intra-arterial infusion site (preferably by administration into a side branch such as is possible in the superior mesenteric artery preparation or the hepatic arterial preparation) [184] (Figure 8.5).

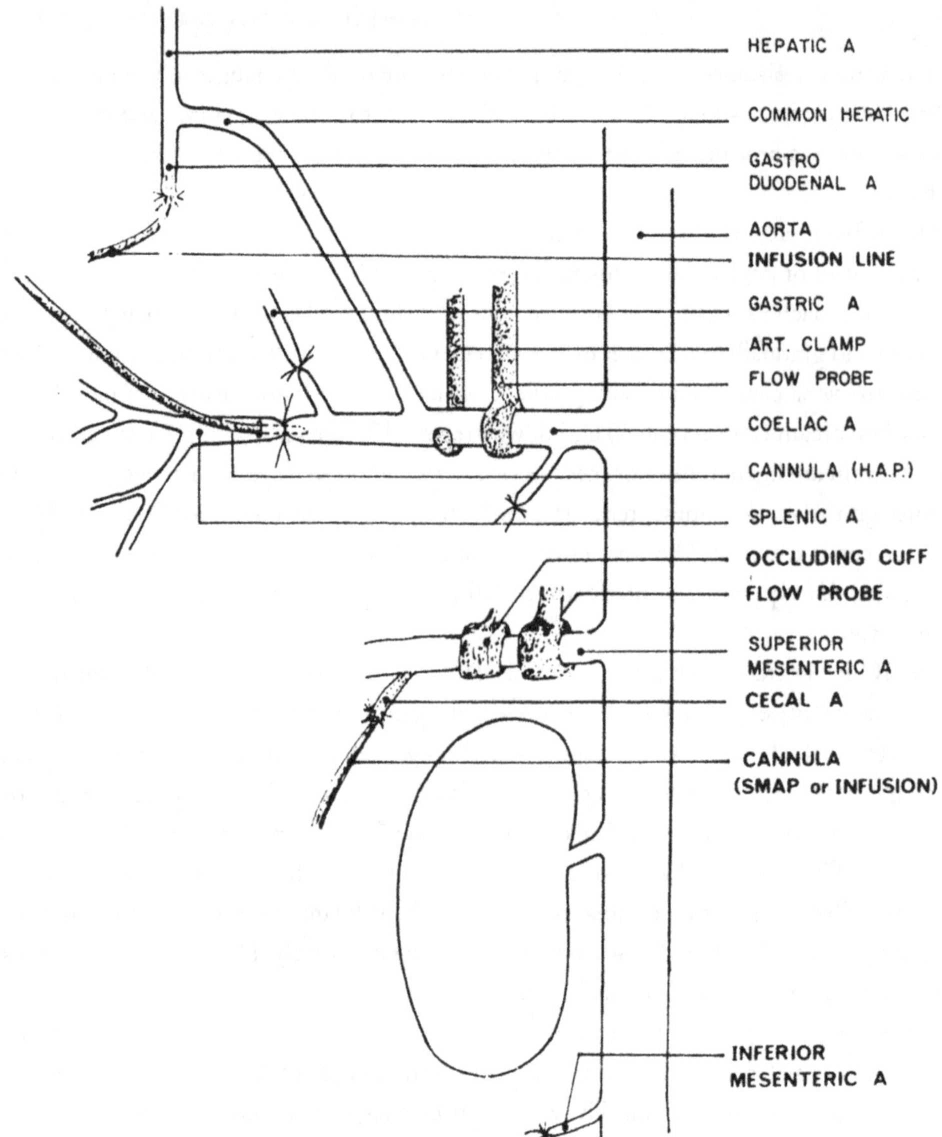

FIGURE 8.5: Preparation used to study pharmacological intervention with hepatic arterial buffer response. Portal flow to liver is made equal to superior mesenteric arterial flow by removal of the spleen, occlusion of the gastroduodenal artery, inferior mesenteric artery, and gastric artery. Portal flow is reduced using an occluding cuff on the artery. Reduced portal flow causes elevated arterial pressure, which is held steady in the hepatic artery using a micrometer-controlled screw clamp and pressure monitored downstream from the clamp via the splenic artery. Because of anastomotic connections with superior mesenteric blood supply, all areas of the gut receive blood and appear normal. A, artery; SMAP, superior mesenteric arterial pressure; HAP, hepatic arterial pressure. Reproduced from Lautt WW, Legare DJ, d'Almeida MS. Adenosine as putative regulator of hepatic arterial flow (the buffer response). *Am J Physiol* 248: pp. H331–H338, 1985. (This figure from publication Am J Physiol is reproduced with permission from publisher).

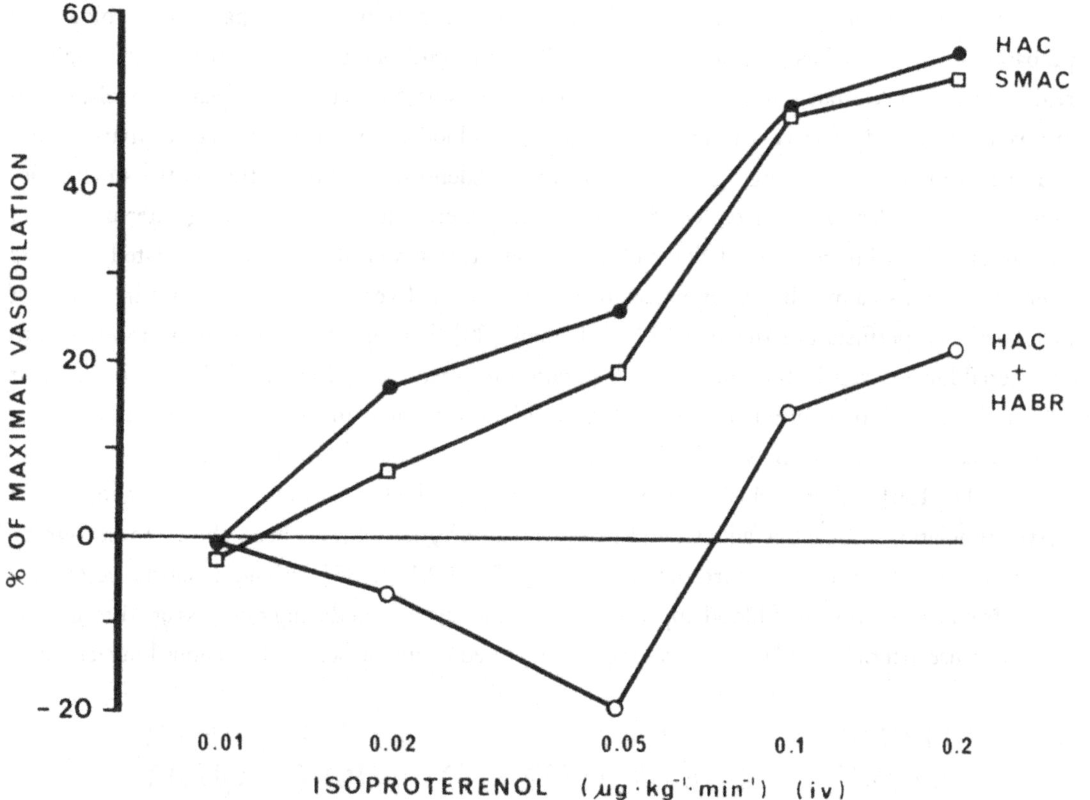

FIGURE 8.6: Responses in one cat to intravenous isoproterenol infusion showing the change in superior mesenteric arterial conductance (SMAC) and hepatic arterial conductance (HAC) expressed as a percentage of the maximal vasodilator responses seen with intra-arterial isoproterenol infusions for each artery. The response of the HA is shown first with SMA flow allowed to rise and thus represents the net effect of the constriction induced by the hepatic arterial buffer response (HABR) and the direct dilator effect of the drug (HAC + HABR). When SMA flow was returned, by clamp, to preinfusion control levels, thus eliminating any HABR constrictor stimulus, the HA dilated to similar levels as the SMA (HAC). The HA response was suppressed by more than 50% through action of the HABR and, at some doses, actually led to a net vasoconstriction. When drugs are administered intravenously, the effect on the hepatic artery is conflicted as the direct drug effect will be countered by the effect of the drug on portal blood flow and the HABR. Reproduced with permission from Lautt WW, d'Almeida MS, McQuaker J, D'Aleo L. Impact of the hepatic arterial buffer response on splanchnic vascular responses to intravenous adenosine, isoproterenol, and glucagon. *Can J Physiol Pharmacol* 66(6): pp. 807–813, 1988. © 2008 NRC Canada or its licensors. (This figure from publication Can J Physiol Pharmacol is reproduced with permission from publisher NRC Canada).

When studying physiological vascular responses, attention must be paid to the biological preparation and related assumptions. For example, many studies attempt to provide data implying regional vascular resistance responses through studies of isolated aorta or other large vessels, which are well recognized to serve a function of conducting blood at low resistance rather than serving as resistance vessels regulating regional blood supply. Adenosine is more active on the small resistance vessels and less on the large conducting vessels, whereas nitric oxide has a greater impact on the larger vessels. Greenway and Stark [122] concluded that vascular data from isolated perfused liver preparations cannot be extrapolated to the intact liver. Even studies carried out in situ must avoid artificial perfusion of the arterial bed. Folkow [80] first reported that a pump inserted into an arterial long-circuit in the hind-limb preparation resulted in significant dilation of the vascular bed, and even minor manipulation of the arterial blood resulted in decreased vascular resistance and reduced reactivity of the vessels. These conclusions have been confirmed and extended to the intestine [73,156] and liver [117]. The response of vascular beds to stimuli, including sympathetic nerve stimulation, vasoactive agents, and myogenic autoregulation, has been shown to be significantly altered in the presence of arterial long-circuits [73,102,117,135]. Whatever substance(s) is released from the manipulated blood appears to be completely cleared during one passage through the lungs because arterial vascular responses appear unaltered in the presence of a venous long-circuit.

8.4 EFFECT OF VASOACTIVE DRUGS ON THE HEPATIC ARTERY WHEN ADMINISTERED INTRAVENOUSLY

The splanchnic vascular interactions are too complex to allow pharmacological responses to be accurately determined using bolus injections of drugs. The hepatic arterial buffer response is a powerful regulator of hepatic circulation. In a situation where portal blood flow increases, either as a result of a normal physiological response or in response to a drug, the increase in portal flow to the liver will activate the buffer response. The competitive effects of the direct vasodilator effect on the hepatic artery and the indirect effect through the change in portal blood flow can result in the hepatic artery showing vasoconstriction to intravenously administered vasodilator drugs [208] (Figure 8.6). When these same drugs are administered directly to the hepatic artery, they demonstrate vasodilator properties typical of other arteries.

• • • •

CHAPTER 9

Nitric Oxide

Nitric oxide (NO) plays several crucial roles in the homeostatic functions of the liver. NO is a gas that has an extremely limited life span unless it is bound to carriers, which it is superbly adapted to do. NO has many different and contrasting functions in different tissues, even within the liver. For example, NO inhibits norepinephrine release from sympathetic nerves supplying the intestine, whereas it does not have presynaptic effects in the hepatic nerves but rather modulates hepatic sympathetic nerve activity through actions at the postsynaptic sites. The vasoconstrictor and glycogenolytic effects of the hepatic sympathetic nerves are affected dramatically differently by NO, which inhibits the vascular response but potentiates the metabolic response. NO also plays a pivotal role in glucose metabolism by virtue of being a mediator of hepatic parasympathetic nerve action in the liver, resulting in NO-dependent regulation of peripheral insulin resistance. This is an entirely separate story, but one which places NO as a major homeostatic regulator whose dysfunction links to prediabetic insulin resistance, obesity, the metabolic syndrome, and diabetes. The role of NO in these processes is outside of the purview of this chapter but has been reviewed recently [202].

9.1 SHEAR STRESS

NO, a potent vasodilator, does not appear to contribute to the basal vascular tone of either the hepatic artery or the portal vein based on the observation that the NO synthase antagonist, L-NAME, did not result in increased vascular resistance and did not interfere with the hepatic arterial buffer response or autoregulation in the liver. This is in contrast to the intestine, in which NO normally contributes to vascular tone and in which elimination of this endothelial-dependent vasodilator tone results in a compensatory potentiation of other vasodilators [251].

NO serves as a modulator of the sympathetic nerves, but the modulation is tightly regulated by whether the vasoconstriction induced by the sympathetic nerves increases shear stress or not. Shear stress does not occur at the site of vasoconstriction if blood flow declines to the extent that perfusion pressure does not increase. However, if perfusion pressure rises during the vasoconstriction, shear stress will be increased and result in release of NO [251,252]. When hepatic sympathetic nerves are stimulated in isolation, portal venous flow is not reduced but portal venous pressure is elevated, which results in increased shear stress and release of NO, which causes suppression of the nerve-induced vasoconstriction. If portal blood flow is experimentally controlled so that the increase in intrahepatic resistance is compensated for by a mechanical decrease in portal flow, the increased hepatic resistance will not result in shear stress and, therefore, no suppression of hepatic sympathetic nerves.

The mechanism of suppression of the vasoconstriction appears to be postsynaptic because the vasoconstriction induced by sympathetic nerve stimulation and exogenous norepinephrine infusion is inhibited to a similar degree [252]. But the role of the hepatic sympathetic nerves in causing glycogenolysis to supply glucose to extrahepatic tissues is preserved and, in fact, potentiated by NO [271,272]. Portal blood flow is capable of stimulating shear stress-induced release of NO that acts on the hepatic artery. This is demonstrated by the inhibitory effect of increased portal shear stress on the hepatic artery, which is eliminated by blockade of NO synthase [252].

As discussed in the chapter on hepatic neurogenic vascular responses (Chapter 11), the phenomenon of vascular escape from vasoconstriction is largely controlled by endogenous release of NO, almost certainly released in response to increased shear stress in both the hepatic artery and portal venous resistance sites. Inhibition of NO synthase, using L-NAME, resulted in almost complete elimination of the hepatic arterial vascular escape in response to continued norepinephrine infusion, and administration of L-arginine, the source of endogenous NO, largely restored the degree of vascular escape [271]. The phenomenon of vascular escape is also seen in the presinusoidal portal resistance vessels, but the mechanism has not been specifically linked to NO release, although shear stress in the portal circulation is capable of releasing NO that acts on the hepatic artery. Vascular escape in the portal vein occurs in response to nerve stimulation but not norepinephrine infusion, in contrast to the hepatic artery, which escapes from both constrictor stimuli. In this regard, the portal venous regulation by sympathetic nerves appears more similar to the superior mesenteric artery than the hepatic artery. It would appear that the age-old quest for the mechanism of arterial vascular escape may be related to shear stress-induced release of NO.

Stellate cells, which have also been previously referred to as fat-storing cells, Ito cells, or hepatic perisinusoidal lipocytes, reside in the perisinusoidal space of Disse and can undergo reversible contraction [160,320]. The speculation that stellate cell activation can result in alterations of sinusoidal pressure is consistent with their histological and anatomical features because they express smooth muscle chemistry that allows these cells to contract [319]. Stellate cells wrap around the exterior of hepatic sinusoidal endothelial cells and, if activated, can theoretically increase sinusoidal vascular pressure. Although the normal functional relevance of the stellate cells to the hepatic vasculature is not clear, it has been suggested that a balance between dilators and constrictors becomes disrupted in conditions of reperfusion injury or endotoxic shock and that the balance favors stellate cell constriction, thus leading to sinusoidal regional blood flow heterogeneity [54]. A connection between NO and the regulation of this sinusoidal tonus via regulation of stellate cells has been suggested based on experiments in which sodium nitroprusside induced stellate cell relaxation and caused partial degradation of actin stressed fibers in these cells [160]. The type of NO synthase that is present in the stellate cells seems to be the inducible form [130].

Kupffer cells are liver-specific macrophages that reside in the sinusoidal lumen, where they are exposed to the systemic circulation via the hepatic artery and to the splanchnic circulation via the portal vein. They constitute approximately 80% of the total population of macrophages in the body [291]. Kupffer cells synthesize inducible NO synthase, producing large amounts of NO [348]. The negative effects of NO may mostly be directed toward the extensive production of NO by the Kupffer cells, which produce NO as a major early response to many toxic actions on the liver. However, overexpression of inducible NO synthase may contribute to liver damage [376]. The damaging effects of NO have been attributed to a cooperative action with superoxide, yielding the peroxynitrite anion ($ONNO^-$). Peroxynitrite, known to oxidize sulfhydryls and to generate products indicative of hydroxyl radical reaction with deoxyribose and dimethyl sulfoxide, induces lipid peroxidation [18,304].

Kupffer cells could play a role in the regulation of the microvascular flow by protruding their pseudopods toward the vascular space in response to endotoxin administration in vivo [260,289]. Moreover, the NO produced by Kupffer cells could affect microvascular flow because of the strong vasorelaxant properties acting on other cells.

NO in the hepatic microvasculature serves a broader protective role. Intravascular administration of melanoma cells in the liver induces a rapid local release of NO that causes apoptosis of the melanoma cells and inhibits their subsequent development into hepatic metastases. Two thirds of the administered cells were arrested in the sinusoids with the remaining caught in the terminal portal venules. Apoptosis was twofold greater in the sinusoids. Inhibition of NO synthase, using L-NAME, blocked the NO burst and inhibited apoptosis of the cancer cells in the sinusoid by 77%, whereas the tumor cell apoptosis in the terminal portal venules was not changed [374].

Kupffer cells exert their NO-producing capacity to depolarize mitochondrial membrane potential in cultured colon carcinoma cells, suggesting a physiological relevance of NO as an immune defense mechanism [175]. On the other hand, when inducible NO synthase activity is upregulated by endotoxin in the liver, an increase in intrahepatic NO generation causes mitochondrial dysfunction in the perfused rat liver [176].

Shear stress-induced NO release results in an additional regional protective action in the liver. Increase in shear stress caused by an increase in the portal blood flow-to-liver mass results in alterations of liver mass as a classic homeostatically regulated mechanism. Increased portal flow increases shear stress, which increases NO, which increases hepatocyte proliferation. The increase in liver size and the associated increase in microvascular circulation restore the blood flow/liver mass ratio. Liver mass is regulated according to hepatic blood flow (see Chapter 15).

• • • •

CHAPTER 10

Adenosine

The role of adenosine is discussed in detail in Chapters 5, 8, 11, 13, and 16. In this chapter, only a summary overview is presented.

Adenosine has a direct vasodilator effect on the hepatic artery but not on the portal vein. Adenosine strongly inhibits vasoconstrictors of the hepatic artery including sympathetic nerve stimulation, norepinephrine, and angiotensin but is without similar effect on the portal vein.

Adenosine is produced from a number of sources including breakdown of mRNA and the adenine nucleotides ATP, ADP, and AMP. These sources are related to parenchymal cell metabolism and oxygen consumption. In addition, adenosine is produced from *S*-adenosyl-homocysteine, in an oxygen-independent manner that regulates adenosine concentration in the space of Mall, thus accounting for the mechanism of the hepatic arterial buffer response, hepatic arterial autoregulation, and the hepatorenal reflex.

The hepatic arterial buffer response (Chapter 5) is regulated by adenosine concentrations in the space of Mall acting on the hepatic artery. Adenosine appears to be secreted at a constant rate into the space of Mall by an oxygen-independent mechanism, most likely from breakdown of *S*-adenosyl-homocysteine. The adenosine is secreted at a constant rate and washed out of the space of Mall into the portal vein. A decrease in portal flow washes away less adenosine, which accumulates to dilate the hepatic artery. Adenosine also serves as the mechanism for hepatic arterial autoregulation, which had previously been thought to be myogenic in nature. Similar to the washout hypothesis, explaining the buffer response, an increase in hepatic arterial pressure will result in an increase in hepatic arterial blood flow that will wash away adenosine from the space of Mall, thereby decreasing its concentration in the vicinity of the hepatic arterial resistance vessels and leading to constriction of the hepatic artery.

Adenosine in the space of Mall activates sensory nerves, thereby providing the first example of sensory autonomic nerves detecting regional blood flow (Chapter 13). Increased adenosine concentration, which results from decreased hepatic blood flow, leads to activation of the sensory nerves that, in turn, activate the sympathetic nerves to the kidneys and result in salt and water retention. Although this is a useful mechanism in a healthy liver, to assist with maintaining hepatic blood flow constant, it is counterproductive in the severely diseased livers in the presence of portacaval shunts.

Portacaval shunting of blood away from the liver leads to elevated intrahepatic adenosine and resultant activation of the hepatorenal reflex with retention of fluid. The fluid retention, in the diseased liver, does not result in restoration of hepatic blood flow but rather leads to massive accumulation of fluids, which can destabilize the entire cardiovascular system.

Adenosine can be produced in large amounts from the hepatic parenchymal cells. The hepatic venous efflux of adenosine can be potentiated by blocking adenosine reuptake mechanisms; however, adenosine released into the sinusoidal spaces from parenchymal or other sinusoidal cells does not have access to the resistance inlet vessels because they are anatomically located upstream from the parenchymal cells and are therefore inaccessible to any substances released by the parenchymal cells. Adenosine released from these sites, however, will have access to the hepatic venous resistance and capacitance sites. However, even high levels of adenosine have minor effects on the hepatic capacitance vessels. Adenosine appears to have no impact on the basal capacitance tone but can result in a small degree of suppression of nerve-induced capacitance responses, which is unlikely to play a significant physiological role as a neuromodulator of the capacitance vessels.

Caffeine produces a powerful antagonism of adenosine A_1 receptors and is capable of blocking the hepatorenal reflex through blocking A_1 receptors on the sensory nerves. In contrast, even very high doses of caffeine have a minor effect on A_2 receptors and the vasculature. The small effect of caffeine on the A_2 receptors appears to act in a noncompetitive manner.

• • • •

CHAPTER 11

Hepatic Nerves

The liver is an extremely richly innervated organ. The functions of hepatic nerves have managed to remain ignored even when they are reasonably clearly understood. In what was the first comprehensive review of the hepatic autonomic nervous system [190], I shared the general frustration felt by researchers at that time.

Hepatic nerve is really a quite useless appellation.
Of function it tells nothing, serving only for location.
Location, too, is doubtful: of function we know less.
The state of knowledge of these nerves is in a dreadful mess.

The hepatic sympathetic and parasympathetic nerves play important roles in regulating a wide variety of hepatic functions. The hepatic parasympathetic nerves regulate "insulin sensitivity" in skeletal muscle ("The HISS Hypothesis" reviewed in 202). The sympathetic nerves affect all aspects of the hepatic circulation and the impact on the overall cardiovascular system is considerable. In addition, the liver may well be the richest sensory organ in the body with afferent nerves that play largely unknown integrative roles as sensors of blood temperature, pressure, and the ionic and nutrient content of portal blood. Sensory nerves in the liver indirectly monitor hepatic blood flow by responding to adenosine levels in the space of Mall. Implications of the hepatic sensory system have been suggested for internal homeostasis, including control of the kidneys, as well as behavioral control of feeding and thirst.

11.1 EXTRINSIC NERVE SUPPLY

Previous reviews have summarized the early anatomical studies [190,334]. Sympathetic nerves (T7–10) reach the liver via the celiac plexus and intermingle with parasympathetic nerves in the right and left vagus and perhaps the right phrenic nerve [3,306]. The posterior hepatic plexus ramifies around the bile duct and portal vein and freely communicates with the anterior plexus travelling with the hepatic artery. The anterior plexus receives nerves from the left and right celiac ganglion and the left vagus nerve. The anterior plexus supplies the cystic duct, gallbladder, and the pancreatico-choledochus nerve. It forms a sheath around the hepatic artery, which can be conveniently prepared for stimulation in physiological preparations in most species (Figures 11.1 and 11.2). The posterior plexus derives from the right celiac ganglion and the right vagus [3]. The majority of nerves enter the liver in association with the blood vessels and bile duct, branching and communicating in the connective tissue of the perivascular spaces [355,356]. Some parasympathetic branches from the left vagus may pass directly to the liver outside of the two major plexuses [3]. The hepatic para-

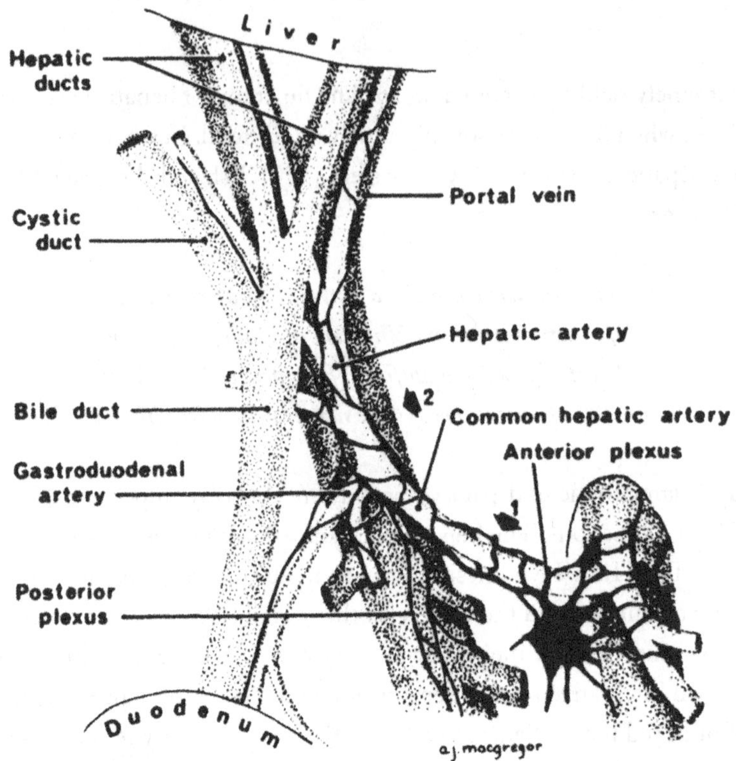

FIGURE 11.1: Ventral view of the anterior hepatic plexus running from the celiac ganglion along the common truck of the hepatic artery, and the posterior hepatic plexus passing to the liver along the portal vessel. The two plexuses freely mingle in the region where the hepatic artery and the portal vein run in close apposition. The biliary tracts are shown leading from the liver at the top of the illustration (the lobes of which are elevated for better access during surgery) to the duodenum at the bottom of the illustration. The arrows show the site at which the anterior plexus (1) can be conveniently isolated for ablation or stimulation and the site, along the portal vessel (2), at which full hepatic denervation can be performed. The latter site cannot be as conveniently used for stimulation since the plexus is very diffuse. Reproduced with permission from Lautt WW. Evaluation of surgical denervation of the liver in cats. *Can J Physiol Pharmacol* 59(9): pp. 1013–1016, 1981. © 2008 NRC Canada or its licensors. (This figure from publication Can J Physiol Pharmacol is reproduced with permission from publisher NRC Canada).

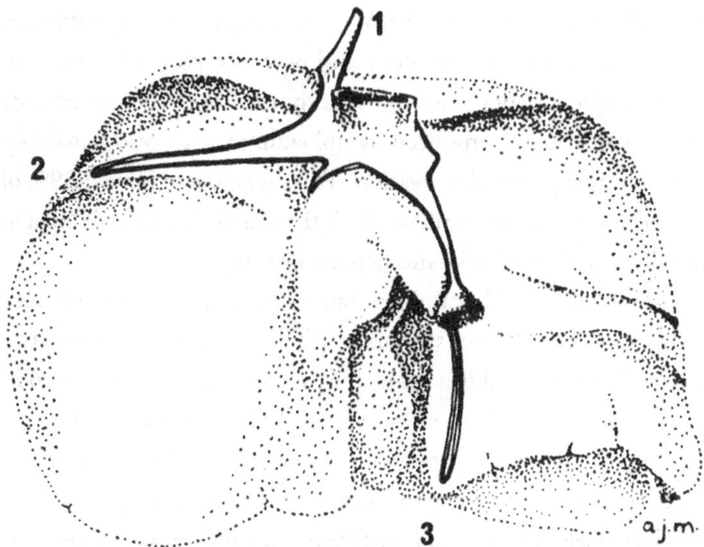

FIGURE 11.2: Dorsal view (visceral surface) of the cat liver showing the location of (1) the falciform ligament, (2) the left triangular ligament, and (3) the right triangular ligament. On several occasions, branches of the vagus could be identified in the left triangular ligament. Fine unidentified nerves are also frequently seen in the falciform ligament. Reproduced with permission from Lautt WW. Evaluation of surgical denervation of the liver in cats. *Can J Physiol Pharmacol* 59(9): pp. 1013–1016, 1981. © 2008 NRC Canada or its licensors. (This figure from publication Can J Physiol Pharmacol is reproduced with permission from publisher NRC Canada).

sympathetic regulation of skeletal muscle glucose uptake can be readily quantified (see a review of the HISS hypothesis [202]) and the response to a meal can be blocked by interrupting the vagus nerve at the neck, at the tiny hepatic branch of the vagus, close to the liver at the anterior plexus, or at the common hepatic artery. All of the vagal nerves in the liver that regulate skeletal muscle metabolism arrive at the liver through the hepatic branch of the vagus nerve. All of the fibers pass along the hepatic anterior plexus, thus allowing complete surgical intervention of the hepatic parasympathetic nerves by denervation at the anterior plexus [183]. The gallbladder and bile ducts and the hepatic parenchymal cells (metabolic control) receive both sympathetic and parasympathetic nerves [306].

The fact that the liver is supplied by two plexuses, and perhaps other nerve branches reaching the liver via other routes, makes surgical denervation difficult and may complicate nerve stimulation experiments. Variability in nerve distribution to the liver has been reported for humans and lower species. A study in the human showed a marked variability in distribution of the branches of the vagus to the stomach and liver in 100 cadavers [345]. They found no constant pattern and concluded that many descriptions in the literature are misleading at best, with many surgical selective vagotomies having resulted in incomplete or erroneous denervations. Variable distribution of sympathetic nerves to the liver through the anterior and posterior plexus has been shown by studying the effects of selective surgical denervation on reflexly mediated sympathetic vasoconstriction in cats [193]. The arterial constriction produced by unloading the carotid baroreceptors was affected from 0% to 100% by anterior plexus denervation. The mean data showed 59% of the constriction being eliminated by cutting the anterior plexus with the remaining 41% being abolished by cutting the posterior plexus along the portal vein and hepatic ligaments.

Comparisons of metabolic and portal resistance responses in the perfused rat liver indicated more marked effects with anterior plexus stimulation. The metabolic responses to stimulation of the anterior and posterior plexus were additive, but this was not true for the hemodynamic responses. The arterial responses were similar with either plexus, but the portal responses were greater with anterior plexus stimulation [90]. In dogs, tyrosine caused release of noradrenaline and induced vascular responses when injected into the hepatic artery but not into the portal vein [88]. The microvascular responses of all vessels visible on the surface of the rat liver were more dramatically affected by stimulation of the nerves around the hepatic artery than around the portal vein [311].

Full denervation can be ensured surgically by tedious and lengthy procedure or, even more difficult, by complete removal of the liver and replacement. For long-term study, both methods result in formation of adhesions [60]. Surgical denervation can be somewhat simplified by applying a few drops of 1% toluidine blue solution to identify nerve fibers [137]. A simpler and more reliable method of producing denervation is demonstrated using phenol (85%) applied on a swab and painted around each of the vessels. This form of denervation produces full functional denervation within 20 min of application [205] that is still complete by 8 weeks post-application with no obvious untoward effects [60]. Both surgical and phenol denervation are nonselective.

Selective sympathectomy of the liver can be produced using intraportal injections of 6-hydroxydopamine. Methods have been developed to demonstrate effectiveness and selectivity of such sympathectomy in the cat [206] and rat [60,207]. Stimulation of the remaining nerve fibers after chronic 6-hydroxydopamine exposure results in a parasympathetically induced reduction in glucose output [235], which is taken as evidence of selectivity. Some authors [4] have preferred to pretreat animals with phentolamine and propranolol to protect against the severe hypertension that was seen with systemic administration of 6-hydroxydopamine. This type of pretreatment does not appear to affect the denervation. The effectiveness of 6-hydroxydopamine is dose-dependent with doses of 50–100 mg/kg, producing low levels of noradrenaline after 1 week but the lower doses resulting in more rapid reinnervation [60]. Although a few researchers (see, e.g., reference [247]) have been scrupulous in their demonstration of functional denervation, most publications that are reported to have produced a denervation offer no evidence of effective denervation. This is especially critical when reporting a lack of effect of a presumed denervation procedure. One advantage of the use of either 6-hydroxydopamine or phenol denervation is that a functional test for sympathectomy is readily available, by electrical stimulation of the anterior plexus and monitoring such responses as elevation in portal venous pressure, decrease in hepatic blood volume, or release of glucose from the liver. Chronic hepatic sympathectomy leads to denervation supersensitivity (Figure 11.3).

The human liver has cholinergic nerve fibers in contact with the intrahepatic branches of the hepatic artery, portal, and hepatic veins [6]. Koo and Liang [170] reported sinusoidal dilation with vagal nerve stimulation and intraportal acetylcholine infusion; however, others have not seen dilator responses to parasympathetic stimulation (reviewed by Greenway and Stark [122]) and topical application of acetylcholine produced constrictions attributable to mast cell degranulation and adrenergic activation [310]. Acetylcholine released from hepatic parasympathetic nerves acts on hepatic parenchymal cells but does not appear to be able to diffuse back upstream from these sites to cause effects on the resistance vessels nor diffuse downstream to affect the capacitance vessels, as

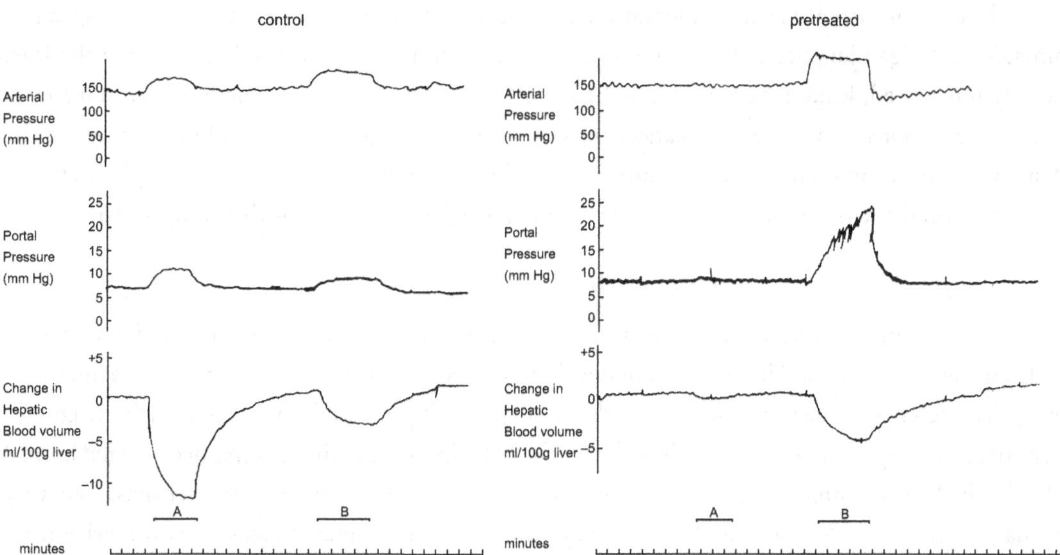

FIGURE 11.3: The response of a normal and a 6-OHDA-pretreated cat to direct stimulation of the hepatic nerves (A) at a frequency of 8 Hz and (B) to intravenous infusion of 1 μg/kg/min of noradrenaline. Pretreatment eliminates the response to direct nerve stimulation, whereas the arterial pressure and portal pressure response to noradrenaline is increased. The volume response to norepinephrine is also increased in spite of a greater opposing portal and arterial pressure. Denervation supersensitivity has occurred. Reproduced from Lautt WW, Cote MG. Functional evaluation of 6-hydroxydopamine-induced sympathectomy in the liver of the cat. *J Pharmacol Exp Ther* 198(3): pp. 562–567, 1976. (This figure from publication J Pharmacol Exp Ther is reproduced with permission from publisher).

parasympathetic stimulation appears without affect and atropine similarly does not result in altered hepatic blood flow, vascular resistances, or blood volume. The parasympathetic system has no apparent direct effect on hepatic vasculature.

11.2 INTRINSIC NERVES

Intrahepatic distribution of nerves shows considerable species variation. The rat and mouse livers have the least extensive adrenergic innervations of hepatocytes, with fibers contacting cells only in zone 1 of the acinus (periportal region). The rat has nerves mainly restricted to the portal space of the hilus, with both cholinergic and adrenergic fibers following the preterminal hepatic artery and, to a lesser extent, the portal venules. The guinea pig and primate livers show more dense innervations where nerves penetrate the acinus right to the central venules [40,266,312]. The human liver shows a rich hepatocyte innervation, and catecholaminergic nerves contact Kupffer cells, endothelial lining cells, and the fat-storing cells of Ito, or stellate cells. Earlier studies and controversies have been reviewed [196].

In those livers with sparse innervation, electrical contact between hepatocytes occurs via gap junctions. The gap junctions appear the most numerous in the areas of the liver that are the least heavily innervated, leading Forssmann and Ito [82] to speculate on their role in electrotonic coupling. The responses to nerve stimulation are similar in the rat, guinea pig, and tree shrew, yet the density of innervations and catecholamine content is remarkably different, the guinea pig liver having 6 times and the tree shrew liver 24 times the noradrenaline content of the rat liver [16].

11.3 DEVELOPMENTAL ASPECTS

Noradrenergic innervation in rats is seen in large portal tracts 1 day after birth and shows final distribution after 5 days. Although nerve development occurs over the same time that acinar heterogeneity of enzyme distribution occurs, the normal development of nerve pattern is only coincident because enzyme heterogeneity develops in the 6-hydroxydopamine-sympathectomized animal [181]. The liver has unique regenerative capacity; after removal of 70% of the liver mass, recovery of mass is essentially complete within 7–10 days. The newly regenerated areas show intensive reinnervation at 10 days; however, if the liver is deprived of either the hepatic arterial or portal venous flow, nerve regeneration is reduced [300]. See Chapter 15 for vascular role in liver regeneration.

11.4 VASCULAR RESPONSES

11.4.1 The Hepatic Artery

Electrical stimulation of the hepatic nerves leads to a marked decrease in hepatic arterial conductance, with blood flow and conductance reaching a minimum level approaching zero at approximately 1–2 min at nerve frequencies of 6–12 Hz in cats and dogs [120]. The nerve frequency required to produce 50% of the maximal conductance response occurs at 2.4 ± 0.9 Hz. The first observations of the response of the hepatic artery to nerve stimulation appear to have been made by Burton-Opitz [42]. Since that time, many people have stimulated the nerves and noted arterial constriction. The pattern of response appears to be of two varieties. The first is typified by the reactions in the dog where the hepatic arterial flow and conductance decreases and remains low for the duration of nerve stimulation [120]. The second pattern is typified by the response in the cat where hepatic arterial flow and conductance decrease but return toward control levels in spite of continued nerve stimulation (Figure 11.4). Vascular escape is a phenomenon not exclusive to cat livers and is reported to occur in a variety of organs and species in response to nerve stimulation and noradrenaline infusion [106,107,191]. Escape from neurogenic vasoconstriction is also seen in presinusoidal portal venous resistance vessels, but the mechanism may be different from that in the artery, as no escape to norepinephrine was produced [219].

This vascular escape from neurogenic stimulation can be distinguished from mere nerve exhaustion. The extent of vascular escape is not dependent on the intensity of initial vasoconstriction

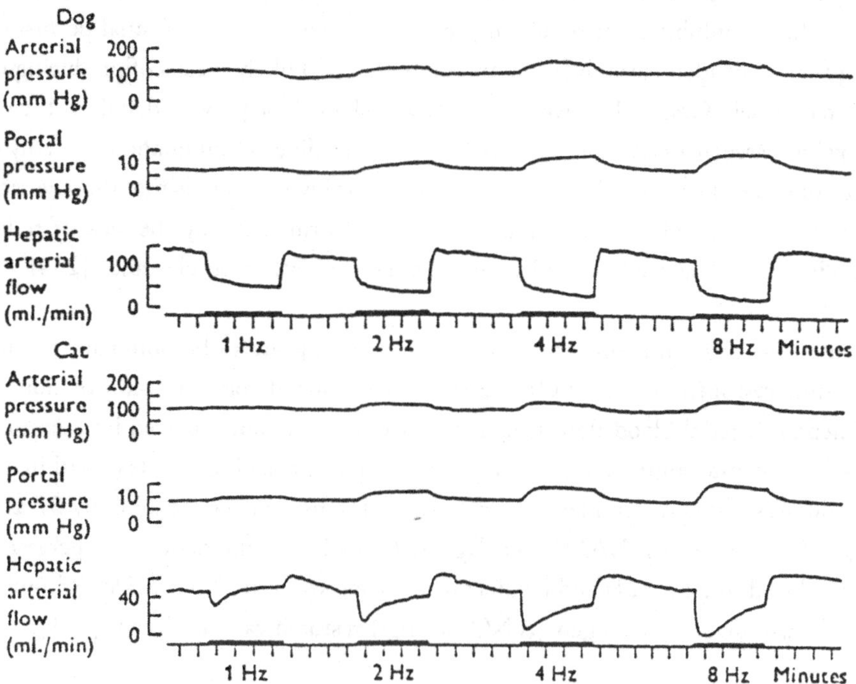

FIGURE 11.4: The arterial pressure, portal pressure, and hepatic arterial flow responses to stimulation of the hepatic nerves in a cat and a dog. Note the occurrence of hepatic arterial vascular escape from nerve stimulation of the flow responses in the cat but not in the dog. Reproduced from Greenway CV and Oshiro G. *J Physiol* 227(2): pp. 487–501. © 1972 by the Cambridge University Press. (This figure from publication J Physiol is reproduced with permission of Cambridge University Press).

[225]. The error in quantification of escape by the use of resistance (escape appears greater from more intense constriction) is of serious consequence if one is attempting to study factors that may modulate the degree of escape.

The distortion of vascular escape produced by calculations of resistance versus conductance has been demonstrated [63]. By the use of resistance, escape will artifactually appear to be reduced by any substance that decreases the initial vasoconstriction (see arguments in Chapter 8 under "Resistance or Conductance?"). Conductance begins to escape from the peak vasoconstriction at 1–2 min of stimulation and reaches a plateau of about an 80% escape by 4–5 min with the time to three quarters of full escape being 3.4 ± 0.5 min [187].

The mechanism of vascular escape from neurogenic constriction has been extensively studied and a variety of possibilities have been excluded. Escape is not due to failure of the nerves because the capacitance vessel constriction is well maintained [123] and hepatic venous resistance remains elevated throughout the stimulation [218]. The escape also occurs in response to infusions of noradrenaline, thereby eliminating the possibility of nerve fatigue being the mechanism. Escape is not modified by β-adrenergic blocking agents; it is not modified by atropine, antihistamines, or prostaglandin synthetase inhibitors. It occurs in preparations with constant arterial perfusion, thereby eliminating flow-dependent accumulation of metabolites as a likely means of producing dilation of the constricted vessels. Comprehensive reviews of vascular escape provide details of the organs and species showing vascular escape and discussion of data leading to elimination of many suggested mechanisms of escape [106,107]. The most likely vascular involvement is that those vessels initially constricted are later partially relaxed. This is supported principally by the lack of redistribution seen with microspheres as well as the lack of interference with drug clearance [234] and oxygen uptake [185].

Vascular escape from neurogenic vasoconstriction appears to be dominantly controlled by shear stress-induced nitric oxide (NO) release. In the control state, infusion of norepinephrine decreased hepatic arterial blood flow, reaching a peak constriction followed by a dramatic return toward baseline, despite continued norepinephrine infusion, reaching a plateau within 3 min. The escape in flow was virtually complete (93 ± 17%) with conductance escaping by 37 ± 6%. After blockade of NO synthase by L-NAME, the degree of initial vasoconstriction was increased and flow escape was reduced to 22 ± 12% and conductance escape was only 7.5 ± 3.3%. Administration of L-arginine, the precursor for endogenous NO, strongly restored vascular escape [271].

11.4.2 Basal Tone

The amount of tonic sympathetic tone on the hepatic arterial resistance vessels is not clear, but several lines of evidence indicate that it is probably absent or very minor in the anesthetized animal.

Acute denervation produced no significant hemodynamic effects in the cat [185], dog [56,284], rat, or rabbit [96].

11.4.3 Reflex Activation

Bilateral carotid occlusion or carotid pressure manipulation leads to reflex constriction of the hepatic artery [49,117,194,363]. Systemic hypercapnia or hypoxia results in reflex hepatic arterial vasoconstriction in dogs that is eliminated or reversed by selective hepatic denervation [255,256]. Whether or not vascular escape occurs to the reflex stimulation is complicated by the fact that such reflex activation is not selective for the liver. In the dog, where escape does not occur to direct stimulation, Mundschau et al. [284] found that reflex activation resulted in an initial constriction, but the hepatic artery showed normal tone by 5 min. It seems highly likely that this sort of response does not represent a true vascular escape but rather represents an interaction between the direct constrictor effect of the hepatic nerve activation and the indirect vasodilator influence of the hepatic arterial buffer response. In this situation the buffer response would be activated secondary to a reflex-induced reduction of blood flow that supplies the portal vein. A reduction in portal venous flow will activate the hepatic arterial buffer response and lead to a dilation of the hepatic artery.

Note that the buffer response is an extremely powerful mechanism that can lead to full vasodilation if portal venous flow decreases sufficiently. The interaction between the buffer response and direct nerve or drug-induced effects on the hepatic artery represents an extremely severe and almost universally ignored complication in hepatic vascular studies. Examples of confusion in the literature are shown by the interactions between the buffer response and intravenously infused vasoactive compounds. Direct acting vasodilators, such as adenosine, isoproterenol, and glucagon, can cause hepatic arterial constriction if the drugs are administered intravenously [208].

11.4.4 Neurovascular Approach in Vivo

Studies requiring quantification of vascular responses require selection of appropriate indices of active vascular responses. Vascular tone in the arteries affects flow more than pressure; therefore, flow should be in the numerator of the equation and arterial vascular tone should be expressed in units of conductance, in preference to units of resistance. This issue is more fully discussed in Chapter 8 where the argument is made with several examples indicating that calculation of arterial vascular tone using conductance (flow ÷ pressure) is preferable to the use of calculated vascular resistance (pressure ÷ flow).

With the realization that the use of vascular conductance would allow classic pharmacodynamic approaches to be used for in vivo studies of intact vascular beds, a powerful research tool was added. The first in vivo study to use arterial conductance responses to calculate maximal vascular responses (R_{max}) and the doses of drugs required to produce 50% of the maximal response (ED_{50}) was reported in 1985 [216], when 8-phenyltheophylline was shown to be a competitive antagonist of adenosine's vasodilator effect on the hepatic artery. In that study the R_{max} and ED_{50} were estimated by linearizing the dose–response curve using the Eadie–Hofstee transformation. Other forms of linearization such as the Lineweaver–Burke method have also been used, but the recent ready availability of the nonlinear regression solving capacities from commercial software packages (e.g., GraphPad, ISI Software) makes the older methods largely of historical interest. We have used the nonlinear regression of the dose–response curves for a variety of studies. Of relevance to this chapter, we have also found that the nerve frequency–response relationship can also be analyzed using this nonlinear regression approach to estimate the maximum response and the nerve frequency required to produce 50% of the maximum response (Hz_{50}). This method thus allows for direct comparisons of responsiveness to nerve stimulation in various disease states and in the presence of various neuromodulators. This classic pharmacodynamic approach, for example, has allowed the demonstration that adenosine is able to produce a dose-related suppression of R_{max} with no effect on the Hz_{50} (classic non-competitive-type antagonism) in the superior mesenteric artery [244]. Many of the technical limitations involved with pharmacological studies (recirculation of drugs, for example) are not of consequence for analysis of the nerve responses. In this regard, it should be noted, however, that a similar pharmacodynamic approach can also be used to analyze frequency–response relationships for hepatic volume responses to nerve stimulation and the pharmacokinetic effect of modulators of this response has been reported using this methodology (see capacitance responses below).

11.4.5 Blood Flow Distribution

Stimulation of hepatic nerves in an isolated perfused liver preparation leads to a marked and well-maintained decrease in oxygen uptake corresponding with a reversible gross heterogeneity of flow, as shown by surface appearance after injection of trypan blue. Approximately 30% of the tissue was estimated to be without flow during stimulation, but flow appeared normal 10 min after cessation of stimulation [155]. Similar heterogeneity was produced in this type of preparation in response to noradrenaline infusion [17]. It was concluded that the mechanism of the reduced oxygen uptake is related to the circulatory changes rather than a metabolic effect [155]. In contrast, flow distribution is not altered in vivo in response to noradrenaline infusion [172] or hepatic nerve stimulation as shown by distribution of radioactive microspheres [119] and unaltered uptake of lidocaine [234] and oxygen [185]. In cats, oxygen uptake decreased transiently but as vascular escape proceeded, oxygen uptake returned to control levels suggesting that some redistribution of flow may have occurred at the onset of stimulation secondary to arterial sphincter-like constriction [188]. Indicator-dilution methodology in dogs confirmed that, despite a 40% reduction in vascular volume (induced reflexly by carotid arterial occlusion), the interstitial space and accessible cell water space were unchanged, indicating that there were no significant zones of no flow [59].

Reports of induction of heterogeneous blood flow with stagnant areas being seen in isolated liver preparations appears to be an artifact of the preparation as nerve-induced heterogeneity is reported for isolated livers but not for livers of the same species in vivo. The dramatic contrast between the isolated, in situ perfused rat liver and the responses of cat and dog livers in vivo reinforces the concerns raised by Greenway and Stark [122] that "conclusions from isolated perfused liver preparations cannot be extrapolated to the intact liver and such preparations have proven of limited value in vascular studies."

11.5 VENOUS RESISTANCE VESSEL RESPONSES TO SYMPATHETIC NERVE STIMULATION

11.5.1 Presinusoidal or Portal Responses

There are literally hundreds of reports in the literature, beginning with Bayliss and Starling in 1894, indicating that stimulation of hepatic sympathetic nerves leads to elevation in portal venous pressure. The specific intrahepatic sites of resistance had not, however, been determined until recently and the introduction of the new index of contractility allows quantification to be done at a more refined level. Previous studies can be found in several reviews [52,114,122,190,196,213,314]. Figure 11.5 shows the response to a 5-min stimulation of the hepatic nerves at a frequency of 8 Hz. The presinusoidal resistance (R_{pv}) increases from insignificant levels in the basal state to account for 50% of the total venous resistance at the peak measured at 2 min. By 5 min, extensive escape of the presinusoidal component had occurred despite the well-maintained constriction of the hepatic venous (hv) site. This amount of vasoconstriction resulted in the presinusoidal pressure gradient rising from an insignificant level (0.2 ± 0.5 mmHg) to account for 39% of the total elevated gradient of 7.2 ± 0.6 mmHg at the peak of the response. After 5 min of stimulation, the pressure gradient had decreased and was only 0.9 ± 0.4 mmHg. In contrast, the hepatic venous pressure gradient remained steady (4.4 ± 0.6 mmHg at 2 min, 4.4 ± 0.8 mmHg at 5 min). This presinusoidal vascular escape is also clearly seen using the index of contractility (Figure 11.5). The mechanism of the vascular escape is likely related to shear stress-induced release of NO (Chapters 5, 6, and 9). It is of interest to note that both the hepatic arterial and the portal venous resistance sites undergo escape, but the hepatic venous resistance and capacitance responses do not escape. The NO-mediated escape in the artery may result from postsynaptic vascular mechanisms, whereas the effect in the portal vein may be mediated by presynaptic suppression of transmitter release as is seen in isolated portal vein segments [36,167].

The index of contractility has a similar qualitative appearance to the calculated responses using resistance as the vascular index (Figure 11.5). However, the resistance calculation grossly

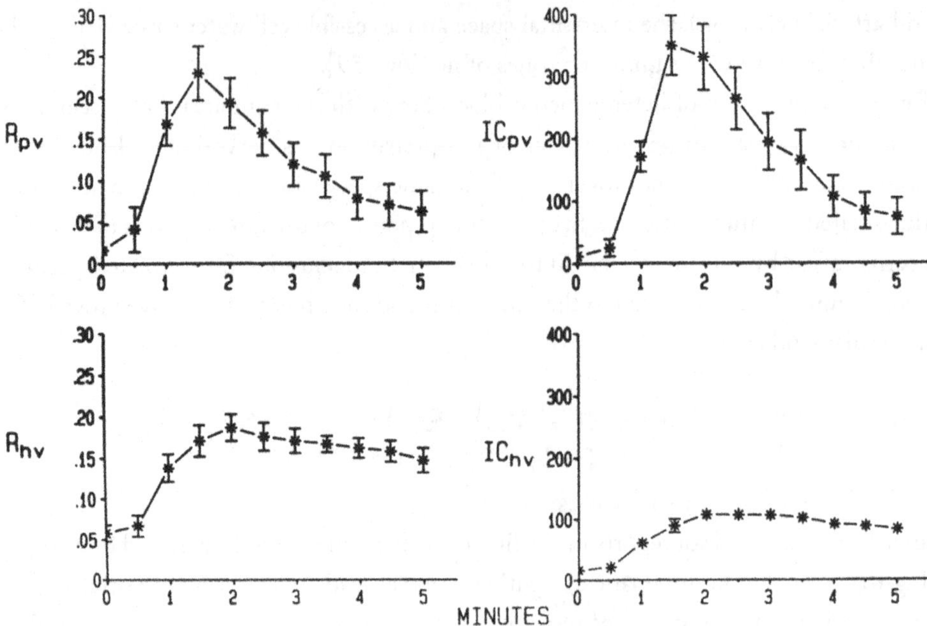

FIGURE 11.5: Mean ± SE responses of portal venous resistance R_{pv} = (PVP – LVP)/(portal flow, mmHg/ml/min/kg) and index of contractility [$IC_{pv} = R_{pv} \times Pd^3_{pv}$ mmHg/ml/min/kg; Pd_{pv} = (PVP – LVP)/2] and hepatic venous resistance [R_{hv} = (LVP – IVCP)/total hepatic flow] and IC [$IC_{hv} = R_{hv} \times Pd^3$; Pd^3 = (LVP + IVCP)/2] to 8 Hz stimulation of the anterior plexus of the hepatic nerve for 5 min measured at 30-s intervals (n = 8). PVP, portal venous pressure. LVP, lobar venous pressure = sinusoidal pressure; IVCP, inferior vena caval pressure. Note the different impression of relative pre- versus postsinusoidal constrictions dependent on use of *R* or IC. Vascular escape occurs at the presinusoidal but not postsinusoidal resistance sites. Reproduced from Lautt WW, Legare DJ. Evaluation of hepatic venous resistance responses using index of contractility (IC). *Am J Physiol* 262: pp. G510–G516, 1992. (This figure from publication Am J Physiol is reproduced with permission from publisher).

quantitatively underestimates the constriction as a result of the interaction between the active vascular constriction and the passive distending effect of the elevated intrahepatic pressures. Figure 11.6 makes this point graphically where the calculated resistance is shown along with the measured distending pressure. Because the index of contractility is not affected by distending pressure [214], the index of contractility would be identical regardless of whether distending pressure had remained steady or was elevated. Therefore, vascular resistance can be calculated from the equation, $IC = R \times P_d^3$, for any distending pressure. In Figure 11.6, resistance has been recalculated for the situation where the distending pressure had not changed. That is, the impact of active constriction and the interaction between active and passive forces have been separated. This exercise indicates that the simultaneous elevation of active vascular tone and distending blood pressure led to a much smaller

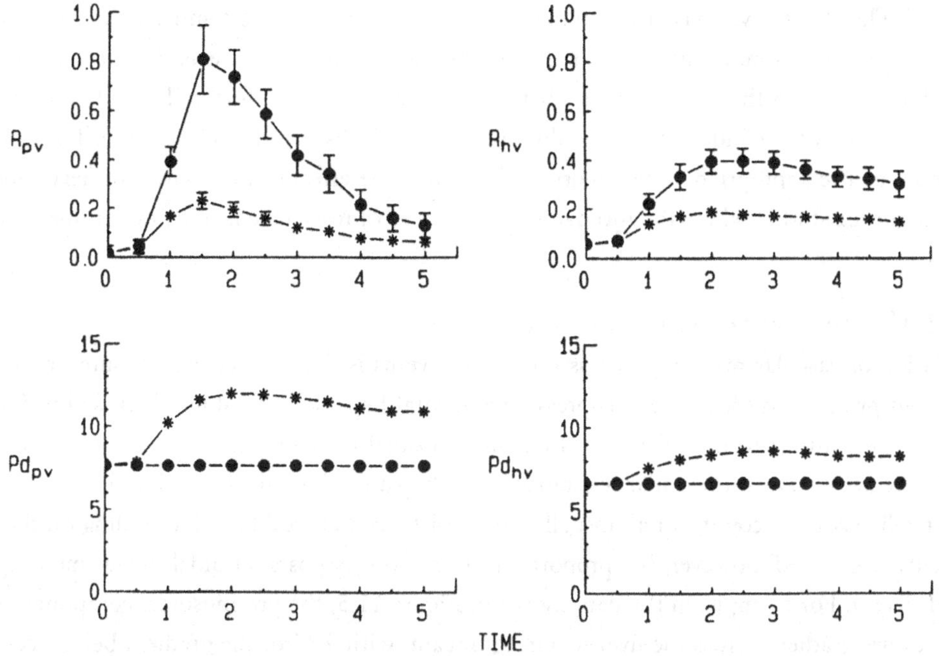

FIGURE 11.6: Resistance of the portal vein (R_{pv}) and hepatic venous segment (R_{hv}) contrasting actual measured resistance (*) during 5 min of nerve stimulation with calculated resistance (•) that would have been determined if the distending blood pressure (P_d) had not risen from control level. This calculation is based on the theory that the index of contractility (IC) would be unaltered by changes in P_d: IC is independent of passive changes in P_d. The calculated R at constant Pd exactly parallels the IC curve in terms of percentage change. IC and R are equally useful indexes of active vascular tone change only if P_d does not change. Differences in change in IC and R are due to the distending effects of P_d on R. The rise in P_d causes the actively contracting sphincters to passively distend thus minimizing the net change in R. Data are means ± SEM. The calculated R at unaltered P_d are means of individual curves rather than being taken from the mean data. Reproduced from Lautt WW, Legare DJ. Evaluation of hepatic venous resistance responses using index of contractility (IC). *Am J Physiol* 262: pp. G510–G516, 1992. (This figure from publication Am J Physiol is reproduced with permission from publisher).

rise in resistance than would have occurred had the distending pressure not changed. The error that is incurred by using vascular resistance as an index of the active vascular tone is in the range of 400%. The change in resistance will accurately reflect active vascular responses only if the distending pressure at the constricting site does not change; the underestimate in the active contractile response by using calculated vascular resistance is magnified as the change in distending pressure is increased.

The selection of the index of vascular reactions must be made rationally and with a clear understanding of the implications of each index. For therapeutic purposes, dealing with amelioration of portal hypertension, the pressure gradient is probably the most important concern for the physician. The effect of various drugs on the intrahepatic pressures is best understood using vascular resistance because the resistance is both affected by the pressure and affects the pressure, and this index thus focuses on the net interaction between active and passive forces. To study purely active responses, however, the index of contractility must be used (discussed in Chapter 8). Thus, this index would be the appropriate index to use to determine the effect of modulators of nerve-induced responses or for studies where the mechanism of changes of resistance is to be determined.

11.5.2 Hepatic Venous Resistance Responses

Calculation of vascular resistance across the hepatic veins is done using the pressure gradient lobar venous pressure to central venous pressure and total hepatic blood flow (discussed in Chapter 6). This is in contrast to the calculations for presinusoidal resistance that uses only portal venous flow. In the basal state, the vascular resistance of the hepatic veins, which is primarily localized to a sphincter-like zone, accounts for almost all of the resistance to blood flow. Depending on the index of vascular tone used, however, the proportion of pre- versus postsinusoidal involvement appears quite different. For example, in the data shown in Figure 11.5, the presinusoidal component of the total pressure gradient across the liver was insignificant, with 90% of the gradient being accounted for by the hepatic veins. The hepatic veins accounted for approximately 78% of the total calculated resistance but only approximately 57% of the calculated index of contractility. Thus, the higher distending pressure at the presinusoidal site produces a passive distention of the presinusoidal vessels that gives the impression that there is no significant vascular tone, unless the index of contractility is used. The active vasoconstriction produced by 8-Hz nerve stimulation is shown in Figure 11.5 where the hepatic venous IC rose by 563% from 16.7 to 110.8 IC units. The distending pressure had also increased over this time (from 6.6 to 8.4 mmHg), and this increase in distending pressure produced a passive dilation of the hepatic venous resistance sites. The net effect on resistance was that it increased by only 222% (from 0.058 to 0.187). The underestimate of active vascular tone responses, using resistance in contrast to the IC, is shown in Figure 11.6. Note that the constriction of the hepatic veins is well maintained for the entire 5-min nerve stimulation, in contrast to the very dramatic vascular escape from neurogenic stimulation seen for the presinusoidal portal venous and hepatic arterial sites. As with the hepatic artery and portal venous responses, appropriate selection of the index of vascular responses is important.

11.6 HEPATIC BLOOD VOLUME

The vascular capacitance of the liver is the total blood volume and is 37 ml/100 g liver (8.2 ml/kg body weight) in denervated cat livers at a portal blood pressure of 8 mmHg [121] with somewhat lesser volume in innervated livers. This is equal to approximately 12% of total blood volume, a value similar to that reported for humans [55].

Capacitance is the total volume, which consists of stressed volume and unstressed volume. Stressed volume is determined by vascular compliance and the intrahepatic blood pressure. Compliance refers to the distensibility of the vascular bed and is defined as the change in volume per unit change in distending blood pressure. Many publications incorrectly interchange the terms compliance and capacitance. There is a linear relationship between distending pressure and volume over a wide physiological range; the slope of the line is the compliance. Compliance is calculated as the change in unit volume per change in unit pressure. Compliance is approximately 2.8 ml/mmHg (per 100 g tissue or 0.6 per kg body weight), and at a pressure of 8 mmHg, the stressed volume is 22 ml/100 g or roughly 60% of total liver capacitance. Extrapolation of the linear pressure–volume curve to zero gives the unstressed volume, a theoretical volume that would exist at zero distending pressure.

11.6.1 Stressed and Unstressed Volume

Liver volume could decrease in response to sympathetic nerve stimulation either by a reduction in compliance or by a reduction in unstressed volume (see also Chapter 4). All available evidence indicates that active venous contraction occurs mainly by a change in unstressed volume [325]. The responses to noradrenaline infusions were studied in the cat liver [121]. Intrahepatic pressure was varied by changing portal flow and by changing outflow pressure, and pressure–volume curves were determined before, during, and after noradrenaline infusion. It was shown that the volume response was entirely due to a reduction in unstressed volume (the pressure–volume curves underwent a parallel shift indicating that compliance was unchanged). When these data are evaluated in conjunction with other data [325], it seems reasonable to conclude that active venoconstriction in the liver and other organs involves a decrease in unstressed volume. Direct support for a similar mechanism of contraction in response to sympathetic nerves has been technically difficult to produce. Bennett et al. [19] reported an apparent decrease in hepatic compliance. They measured the volume response to a single 1-min, 5-mmHg step elevation in venous pressure with the assumption that the pressure was transmitted similarly to the capacitance vessels in control and during nerve stimulation. This is the same assumption that Clive Greenway and I previously erroneously made for compliance measurements [212]. Transmission of a rise in hepatic outflow pressure beyond the hepatic venous resistance sites is dependent on venous resistance and is reduced by nerve stimulation [223] (also see Figure 6.6). This reduced pressure transmission would create an apparent reduced compliance. Figure 11.7 shows the progressive increase in transmission of central venous pressure changes across the venous resistance site as the raised pressure passively dilates the resistance vessels. If the venous

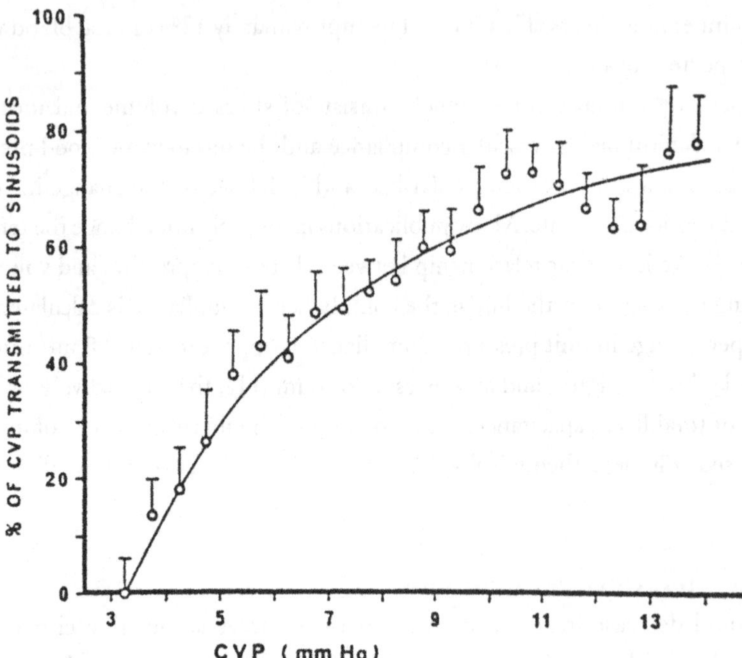

FIGURE 11.7: Proportion of increase in central venous pressure (CVP) transmitted to sinusoidal pressure. A data point at, for example, 5.75-mmHg CVP represents the percentage of pressure rise from 5.5 to 6.0 mmHg CVP that was transferred to the sinusoids. Note that even very small elevations in CVP are partially transmitted upstream to the sinusoids. The percentage of transmission for small elevations in central venous pressure is low but rises as the distending pressure of the central venous pressure leads to distention of the hepatic sphincter and a resultant decrease in sphincter resistance. Active vasoconstriction affects this relationship by reducing the percent transmission at each point. Reprinted from Lautt WW, Greenway CV, Legare DJ. Effect of hepatic nerves, norepinephrine, angiotensin, elevated central venous pressure on postsinusoidal resistance sites and intrahepatic pressures in cats. *Microvasc Res*, vol. 33, no. 1, p. 57. © 1987 by Academic Press. (This figure from publication Microvasc Res is reproduced with permission of publisher Academic Press).

compliance is calculated by dividing the change in volume by the change in blood pressure, the use of the hepatic venous pressure results in a greater apparent compliance for the higher pressure elevation. If, however, one makes the assumption that the pressure gradient across the liver is mainly due to postsinusoidal resistance and uses portal pressure as the distending pressure, the compliance is the same for both degrees of pressure elevation. This study is compatible with the notion that virtually all of the venous resistance is postsinusoidal; that the stressed volume is primarily proximal to the hepatic veins and that portal pressure is a better estimate of the sinusoidal blood pressure than is hepatic venous pressure. The use of hepatic venous pressure will underestimate hepatic compliance and make it appear to be nonlinearly related to pressure.

The concept of venoconstriction as a change in unstressed volume rather than a change in venous compliance (Chapter 4) has interesting physiological consequences [114]. If venoconstriction occurred by a decrease in compliance, the amount of blood actively mobilized would become progressively smaller as the intrahepatic pressure decreased. This would mean that in those situations where mobilization of blood was most vital, for example, during hemorrhage, the amount that could be mobilized by the sympathetic nervous system would become smaller as the hemodynamic status deteriorated and liver volume passively decreased. In contrast, when venoconstriction causes a change in unstressed volume with no change in compliance, the amount of blood mobilized by the sympathetic nervous system is independent of intrahepatic pressure. In this case, the volume mobilized actively by the change in unstressed volume and passively by the decline in intrahepatic pressure is additive. This would give the animal the best chance of survival (Figure 14.1, Chapter 14).

Although the net capacitance response of the liver superficially appears quite straightforward, the interaction between stressed and unstressed volume during sympathetic nerve-induced contraction becomes quite complex. Figure 4.4 in Chapter 4 shows conceptually the interactions that probably occur during nerve stimulation. Nerve stimulation does not likely alter compliance but, because of constriction of the hepatic venous sphincters, intrahepatic pressure increases, thus leading to an increase in the stressed volume. Unstressed volume is actively decreased, with the net effect being a reduction in total capacitance but an increase in the stressed volume. This also would appear to have survival advantage in permitting further additional reduction in capacitance that might occur secondary to passive hemodynamic reactions, such as the reduction in portal pressure that would occur with the reduction in cardiac output during hemorrhage.

Unstressed volume is hemodynamically inactive, but when venoconstriction changes unstressed volume into stressed volume, it is equivalent to a transfusion of a significant amount of blood. Sympathetic nerve stimulation can expel up to 50% of the total blood volume of the liver within 90 s and thus provide an effective transfusion equal to approximately 6% of the total blood volume of the body.

11.6.2 Responses to Direct Nerve Stimulation

Early studies have previously been reviewed [114,122]. Cats and dogs show frequency–response relationships that are very similar. Maximal expulsion of blood volume occurs between 6 and 8 Hz with maximum volumes of expulsion being equivalent to roughly 50% of the total blood volume [120]. The nerve responses are mediated by α_2 adrenoreceptors [339]. The use of nonlinear regression of the frequency–response curves allow quantitative assessment of the maximal response (R_{max}) and the nerve frequency required to produce 50% of this response (Hz_{50}). The use of such dynamic parameters allows for well-defined quantification that can be used to assess the impact of neuromodulators or diseased states on the capacitance responses. If a diseased liver, for example, had a reduced responsiveness to nerve stimulation tested at only one frequency, it would be impossible to determine whether the sensitivity to the stimulation was decreased or whether the overall ability of the liver to develop a capacitance response was diminished. Figure 11.8 shows nonlinear regression best fit to the frequency–response curve indicating a Hz_{50} of 3.2 ± 1.0. A separate series of cats reported in the same study had a Hz_{50} of 1.7 ± 0.3 [231].

These values are in the same range as those that can be determined by visual inspection of frequency–response curves for dogs and cats [120] where the Hz_{50} is in the range of 2 Hz. This is in the same range as the responsiveness of the hepatic arterial resistance vessels (2.4 ± 0.9; see Figure 11.9). The Hz_{50} for volume responses of the large, extrahepatic portal vein is 3.4 Hz [357]. Thus, the resistance vessel responses and capacitance responses appear to show similar intensity of constriction at each nerve frequency. This is in contrast to previous implications that indicated that capacitance responses reach a greater effect at comparable nerve frequencies relative to the maximum response that is seen with the resistance vessels. The impression of a less sensitive capacitance

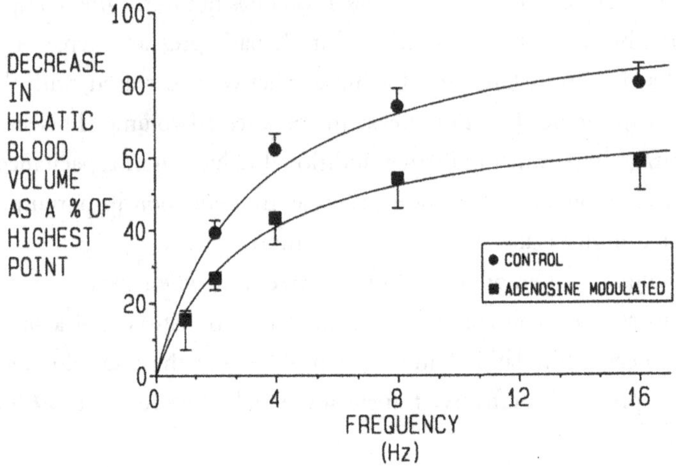

FIGURE 11.8: Standardized, pooled frequency–response curve in control state and during intrahepatic arterial infusion of adenosine. The responses were standardized by expressing each response as a percentage of the largest response obtained in either the control or test curve. The Hz_{50} in the control curve (3.2 ± 1.0 Hz) was not different from that during adenosine infusion (3.1 ± 0.5 Hz). The R_{max} was lower in the adenosine group (100.5 ± 10.7% for control, 72.6 ± 4.3% during adenosine). Adenosine acts as a classic noncompetitive inhibitor of nerve-induced capacitance responses. Reproduced with permission from Lautt WW, Schafer J, Legare DJ. Effect of adenosine and glucagon on hepatic blood volume responses to sympathetic nerves. *Can J Physiol Pharmacol* 69: pp. 43–48, 1991. © 2008 NRC Canada or its licensors. (This figure from publication Can J Physiol Pharmacol is reproduced with permission from publisher NRC Canada).

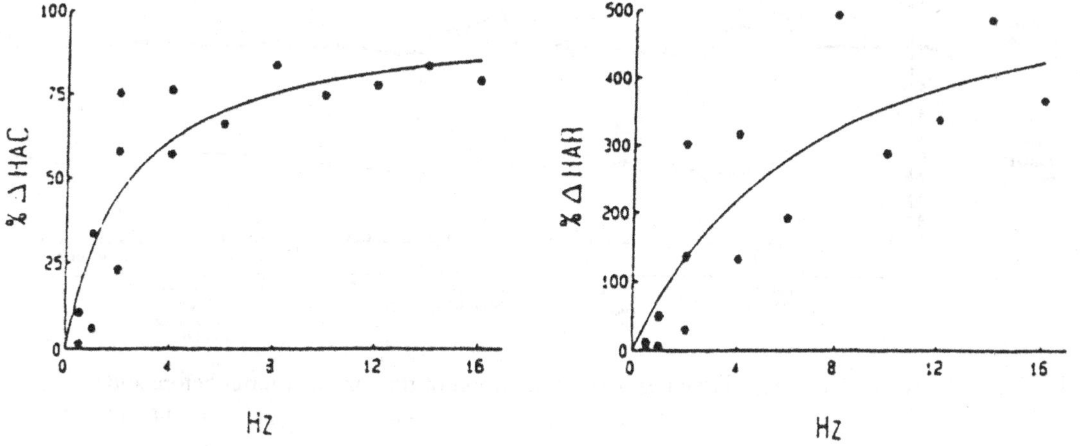

FIGURE 11.9: The peak vasoconstrictions to hepatic nerve stimulation expressed as percent change in hepatic arterial conductance (HAC; note that this is negative, indicating constriction) and resistance (HAR) in one cat. The dynamic parameters of the frequency–response curve are based on nonlinear regression to estimate the maximal response (R_{max}) and the nerve frequency required to produce 50% of the R_{max} (Hz_{50}). The line is best fit to the rectangular hyperbolic curve for data based on HAC and HAR. The HAC is the appropriate index. R_{max} is 96.74 ± 12.13 (12.5%) for %ΔHAC and 610.70 ± 193.30 (31.7%) for %ΔHAR. Hz_{50} is 2.35 ± 0.93 (39.6%) for %ΔHAC and 7.18 ± 4.87 (67.9%) for %ΔHAR. SE is 0.779 for %ΔHAC and 0.718 for %ΔHAR. Reproduced from Lautt WW. Hepatic circulation. In: *Nervous Control of Blood Vessels*, Chapter 13, Figure 13.3, p. 472, 1996. (Figure 13.3 from publication Nervous Control of Blood Vessels is reproduced with permission from publisher Taylor & Francis).

response is probably because the responses were calculated according to vascular resistance rather than conductance. This interpretation is supported by comparing the Hz_{50} based on conductance (2.4 Hz) to that estimated from calculated resistance (7.2 Hz) shown in Figure 11.9, Harwood Academic Publishers GmbH, London, UK.

11.6.3 Reflex Control of Hepatic Capacitance

Reflex control of the venous system has been previously reviewed [325]. Despite the large and clearly demonstrable capacitance responses to direct electrical stimulation of the nerves, physiological involvement of reflex control has been more difficult to convincingly demonstrate (Figure 11.10). The problem is partly related to the fact that most reflexes produce complex interacting responses in many vascular beds. In addition, it is clear that to differentiate direct nerve effects from indirect effects on stressed volume, secondary to reduced splanchnic blood flow, it is necessary to monitor intrahepatic pressures. Perhaps the most severe limitation is that hepatic capacitance responses cannot yet be studied in conscious animals. A number of observations indicate that the

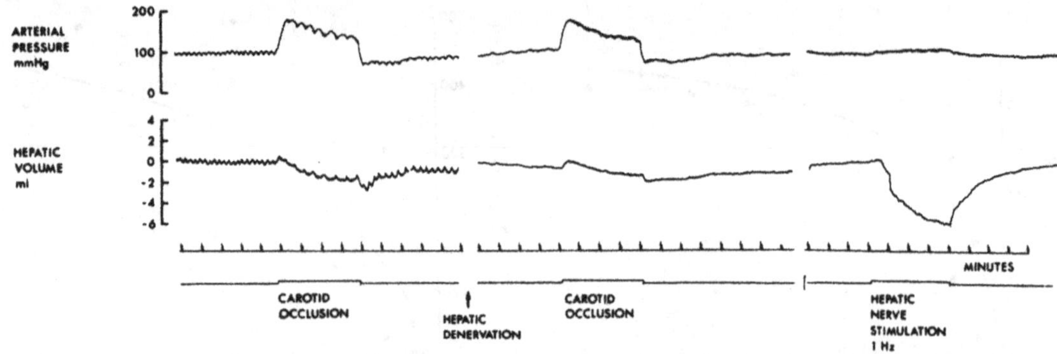

FIGURE 11.10: The hepatic volume response to occlusion of the carotid arteries before and after section of the hepatic nerves and to stimulation of the hepatic nerves at a frequency of 1 Hz. Cat, 3.0 kg; liver weight, 101 g. The hepatic blood volume in this cat was 24 ml. Reproduced with permission from Lautt WW, Greenway CV. Hepatic capacitance vessel responses to bilateral carotid occlusion in anesthetized cats. *Can J Physiol Pharmacol* 50(3): pp. 244–247, 1972. © 2008 NRC Canada or its licensors. (This figure from publication Can J Physiol Pharmacol is reproduced with permission from publisher NRC Canada).

presence of anesthesia blunts the central nervous mechanisms that control the venous system, including the liver [114].

Bilateral carotid occlusion had no effect on hepatic blood volume in cats [194,211,248]. Lack of effect was also confirmed using perfusion of the carotid vessels to manipulate pressure at the carotid sinus baroreceptors [194,248]. The arterial baroreceptors did, however, produce changes in hepatic arterial and other resistance vessels. Either reflex responses of the venous system are blocked by anesthesia or the major reflex mechanisms have not yet been discovered. However, responses to hemorrhage have also shown no evidence of active reflex hepatic venous responses [108,204]. The entire large capacitance response to hemorrhage could be accounted for by the effects of reduced blood flow acting through changes in stressed volume [108].

A variety of studies in dogs consistently demonstrated reductions in hepatic and splanchnic volume on reducing blood pressure at the carotid sinus. Carneiro and Donald [49] found a carotid sinus pressure-related hepatic volume response that led to about a 25% reduction of hepatic volume at sinus pressure reduction of 80 mmHg. Portal and arterial resistance rose at the same time. Because portal pressure rose significantly, the reduced volume would likely represent reduced unstressed volume. Several related studies report splanchnic volume responses, which do not differentiate liver from the other organs; portal pressure is often and intrahepatic pressure is never measured, so mechanisms of volume response are unclear. This work has been reviewed [72,114]. Cousineau et al. [58] used indicator dilution techniques to measure hepatic blood volume in response to bilateral carotid occlusion. Although hepatic blood flow was not altered by this maneuver, hepatic blood volume decreased by 40%. Mild acidosis (pH 7.2) prevented the carotid occlusion-induced hepatic capacitance effect in dogs [99], but it is not clear if this may have been because systemic acidosis (pH 6.9) had already produced active capacitance constriction [327].

Cerebral ischemia led to a neurogenic reduction of volume of the portal vein by up to 26% in rabbits [357] that was equivalent to the volume response produced by 10-Hz transmural electrical field stimulation. Frequency–response curves were presented and indicate a Hz_{50} of approximately 3.4 Hz. The methodology of using in vivo plethysmography on a segment of portal vein was a novel approach for studying venous responses.

The role of the venous system in cardiovascular reflexes is controversial and has been discussed [112,114,325]. There is a large body of circumstantial evidence in favor of reflex control of the venous system and of the splanchnic and hepatic venous beds in particular, but studies directed to unraveling the reflex control of these capacitance beds are inconclusive.

11.7 HEPATIC FLUID EXCHANGE

The major physiological variable controlling net fluid exchange in the liver is sinusoidal hydrostatic pressure. Elevation of hepatic venous pressure results in partial transmission of this pressure to the sinusoid and produces a constant filtration rate that is maintained for at least 6 h, provided plasma volume is maintained by intravenous replacement of filtered fluids [110]. The filtered fluid does not represent pooling in an interstitial fluid compartment because this filtered fluid could be quantitatively collected from a plethysmograph in which the intact liver with ligated lymphatics was contained. The liver thus clearly lacks the protective mechanisms that limit filtration in the intestine and skeletal muscle. Increase in filtration across the porous endothelial cells results in fluid passing into the space of Disse. The hepatic sinusoids are perforated by fenestrae of 1–3 mm [101], and these fenestrae are of the appropriate size to exclude some of the large-molecular-weight lipoproteins from the space of Disse [85]. The physiological importance of the fenestrae in regulation of lipoprotein metabolism is not clear, but there is evidence to suggest that enlargement of the fenestrae size leads to increased cholesterol and lipid uptake by the hepatocytes [84].

Increased intrahepatic pressure results in increased filtration from the blood compartment into hepatic lymphatics and, if the lymphatic system becomes overloaded, filtration across the capsule of the liver. Fluid filters readily across the liver capsule and has approximately the same protein content as the plasma [110]. In diseased livers with elevated intrahepatic pressure, this fluid is referred to as ascites and can expand to as much as 20 liters in volume.

Sympathetic nerve stimulation produces very perplexing responses. Intrahepatic and sinusoidal pressure is clearly elevated (see previous sections); however, increased fluid filtration does not occur and liver volume remains stable for long periods of nerve stimulation [123]. When filtration is

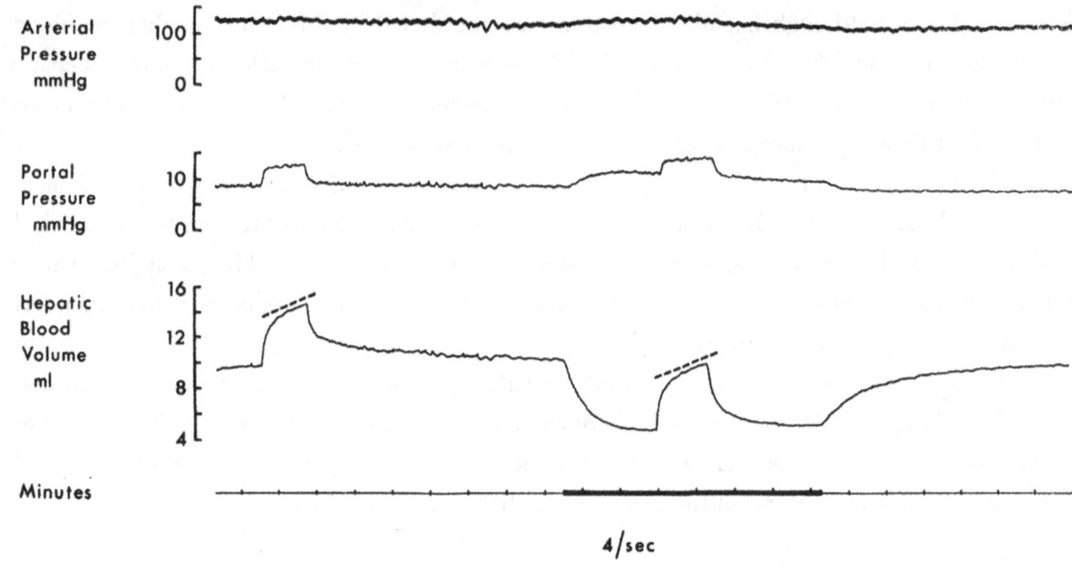

FIGURE 11.11: Determination of the transhepatic filtration rate by raising hepatic venous pressure 4.7 mmHg before and during stimulation of the hepatic nerves at a frequency of 4/s. The broken lines represent the slopes representative of the fluid filtration. The hepatic blood volume was determined shortly after these responses to allow calibration of the capacitance response to nerve stimulation in terms of hepatic blood volume. Reproduced from Greenway CV, Stark RD, Lautt WW. Capacitance responses and fluid exchange in the cat liver during stimulation of the hepatic nerves. *Circ Res* 25: pp. 277–284, 1969. (This figure from publication Circ Res is reproduced with permission from publisher).

induced by elevation of the hepatic venous pressure, nerve stimulation results in no further increase in filtration and, in fact, results in a small decrease (Figure 11.11) [105]. Because the intrahepatic pressure has increased, it appears that the ability of fluid to exit the plasma compartment has been decreased. Similarly, the intrahepatic pressure at which exudates appears is higher in innervated livers than in denervated livers [108]. The nerve-induced reductions in filtration could occur secondary to reduction in size of endothelial fenestrae. Alternatively, the site of restriction could be the small gaps that lead from the space of Disse into the space of Mall from where the lymphatics arise. Reduced fenestrae size is shown to occur in response to noradrenaline and serotonin infusion [378], but effects of nerves have not been studied. Bennett et al. [19], using a constant flow preparation, found that nerve stimulation led to large increases in portal pressure and elevated fluid filtration, as estimated from hematocrit changes. The reason for this difference from Greenway's data is not clear, but the methodologies are dramatically different. Considering the possible physiological and pathophysiological relevance of changes in fenestrae size and the potential for fenestrae to regulate the access of plasma-borne lipoproteins to hepatocyte surfaces, this area requires intensive investigation.

11.8 NORADRENALINE OVERFLOW

Data on noradrenaline overflow into the hepatic venous effluent of the liver are of limited value because of lack of information as to the source of the released noradrenaline (arterial, portal, sinusoidal, venous sites?). In addition, noradrenaline that is released upstream from the region of the arterial or portal venous resistance sites must pass through the sinusoids and past the parenchymal cells, through the hepatic venous sphincters and into the venous effluent before being sampled. Noradrenaline that might be released from the hepatic veins, on the other hand, passes none of these other potential uptake sites.

The importance of the amount of noradrenaline released may be highly species-dependent, with those species such as the rat that have sparse innervations releasing less noradrenaline in contrast to species such as the guinea pig and tree shrew that release substantially more noradrenaline, while producing vascular and metabolic responses that are similar to that of the rat [16]. The rat liver contains few nerves and low levels of noradrenaline compared to the guinea pig and tree shrew livers, but the kinetics of release were six- to sevenfold higher in guinea pigs and tree shrews compared to that in rats. It was estimated from these data that a 5-min stimulation period led to release of approximately 26% of the total hepatic content of noradrenaline in the guinea pig, 6% in the tree shrew, and 22% from the rat liver [16]. Similar data from other species are not available.

Beckh et al. [15] found no correlation of hemodynamic effects of sympathetic nerve stimulation with noradrenaline overflow in the rat liver perfusion preparation, whereas a good correlation is seen in vivo in dogs [382]. The reason for the different results is unclear and may relate to the fact that many in vivo responses are different from results obtained in isolated perfusion systems or the vascular response used as the index might determine the degree of correlation. For example, Yamaguchi and Garceau [382] showed that changes in arterial conductance correlated better than changes in resistance or portal pressure; the hemodynamic response quantified in the rat preparation was the decrease in portal blood flow.

11.9 NEUROTRANSMITTERS AND NEUROMODULATION

The arterial and portal constrictor effects of nerve stimulation are prevented by α-receptor blockade [116,117,189]. On complete blockade by α-adrenergic antagonists, a small β-adrenergic vasodilation can be produced, which is blocked by propranolol [116].

Adenosine is able to suppress the arterial but not portal venous constriction to nerve stimulation, but the response to noradrenaline and angiotensin is equally well suppressed, suggesting postjunctional antagonism [217]. Unpublished observations suggest that such modulation probably does not occur in physiological conditions in the liver. Adenosine receptor blockade does not alter initial vasoconstriction, the extent of escape, or post-stimulatory hyperemia. Glucagon was reported to antagonize the arterial but not portal venous constrictor response to nerve stimulation and the response to noradrenaline and other constrictors in dogs [314,315], but later studies [61,62] failed to confirm this in cats and found that at some doses constriction was actually enhanced to a small extent. Vascular escape [61,62] was suppressed by glucagon, but the doses required to affect the arterial responses to nerve stimulation were well outside of even pathological levels. Whether the different observations are due to the species studies is not clear, and further studies are required to clarify the possible role of glucagon as a neuromodulator. The ecosanoids, discussed later, appear to be a possible family of physiological modulators of hepatic arterial responses to sympathetic nerves. NO, adenosine, hydrogen sulfide, and carbon monoxide are discussed in different sections.

The foregoing discussion of the differences between vascular resistance and index of contractility indicates clearly that modulation studies of the venous resistance sites should be done using the index of contractility. Such studies have not yet been done. However, certain tentative conclusions can be made. The neurotransmitter appears to be noradrenaline, which may or may not be co-released with ATP and other transmitters. The rise in portal pressure in response to sympathetic nerve stimulation can, however, be eliminated by the use of the nonselective α-adrenergic receptor antagonist, phenoxybenzamine [116] or phentolamine [189]. Portal pressure appears to be under the control of both α_1 and α_2 receptors [339]; however, the relative contribution of receptor types at the presinusoidal and postsinusoidal site is completely unknown. The question of modulation of sympathetic nerve responses is extremely important in view of the controversial involvement of the sympathetic nervous system in contributing to elevated portal pressure in the cirrhotic patient [132].

Only a few studies of neuromodulation have been done related to the hepatic capacitance vessels. The blood volume responses to sympathetic nerve stimulation are mediated by α_2 adrenoreceptors [339]. The direct effect of infused noradrenaline was not affected by the calcium channel blocker, nifedipine, but the response to sympathetic nerve stimulation was reduced by 33% and increasing the dose produced no further impairment. Nifedipine may have acted presynaptically to reduce the release of noradrenaline from the sympathetic terminals in the hepatic venous bed [340]. Similarly, bromocryptine, a dopamine (DA2 receptor) agonist that is reported to cause presynaptic inhibition of noradrenaline release, resulted in impaired volume responses to sympathetic nerve stimulation but not portal pressure responses nor responses to direct infusion of noradrenaline [109]. Elevations in intrahepatic pressure also produce selective inhibition of the capacitance responses to nerve stimulation such that volume responses are almost eliminated at intrahepatic pressure of 16 mmHg [105,108], whereas the response to direct infusions of noradrenaline is well maintained [204]. Systemic acidosis (pH 7.2) has been reported to virtually eliminate the hepatic capacitance response to bilateral carotid occlusion in the dog liver. Noradrenaline overflow into the portal and hepatic veins was reduced despite normal arterial pressure responses to the carotid occlusion [99]. Adenosine suppresses the R_{max} but does not alter the Hz50, whereas glucagon increases the Hz50 from 3.4 to 5.6 Hz [214]. Although these compounds showed the ability to modulate the nerve responses, the doses needed were nonphysiological.

Iwai and Jungermann [147] found that inhibitors of eicosanoid synthesis reduced the vasoconstriction induced by ATP and nerve stimulation and noradrenaline infusion. Noradrenaline overflow was not altered, leading to the conclusion that the eicosanoids might be released from parenchymal cells to cause a portion of the vasoconstriction. A role for leukotriene C4 and D4 in this regard has been eliminated [148] and prostaglandin $F_{2\alpha}$ and D2, based largely on their ability to mimic the metabolic and hemodynamic effects of nerve stimulation, are implicated [146]. PGE_2 is released from the liver by nerve stimulation [362]. Hypoxia reduced the constrictor effects of the nerves but not noradrenaline and $PGF_{2\alpha}$. The suggestion was made that oxygen-dependent eicosanoid production is involved with mediating those responses that were suppressed by hypoxia [14]. A similar role for prostaglandin mediation of nerve responses was suggested not only for the rat, where the sparse innervations and low noradrenaline content indicate a need for some sort of cell-to-cell propagation but also in the guinea pig with its much more dense innervations and six-fold higher noradrenaline content and noradrenaline overflow upon stimulation [16]. The oxygen dependence of the eicosanoid involvement in sympathetic nerve responses is potentially very problematic because it clearly appears that sympathetic nerve stimulation in the perfused rat preparation leads to markedly heterogeneous flow, with as much as 30% of the liver sinusoids being unperfused during the stimulation [155], in contrast to the lack of heterogeneity induced in vivo in the cat and dog. It would appear that to supply adequate oxygen to allow normal oxygen uptake using blood cell-free perfusate, hyperphysiological levels of oxygen (75%) and blood flow (4.6 ml/min per gram of tissue) are required [14].

Some degree of presynaptic modulation of transmitter release appears to occur in both the dog [381] and rat [15], α_2 adrenergic receptor antagonism by yohimbine results in a greater transmitter overflow into the hepatic veins of the dog, and clonidine (an α_2 agonist) reduced overflow [381]. Noradrenaline infusion reduced noradrenaline overflow and phentolamine increased overflow in the rat [15]. Thus, there is evidence for α_2 postsynaptic receptors mediating the capacitance responses and α_2 receptors mediating presynaptic noradrenaline release in the resistance vessels. The metabolic, portal, and arterial constrictions in response to nerve stimulation were inhibited by the α_1-adrenergic antagonist, prazosin, in the rat liver [90].

Co-release of neurotransmitters other than noradrenaline is unclear in the liver. The distribution of noradrenergic and neuropeptide Y regulating enzymes suggests co-localization [40]. In both human and rat livers, there is a rich distribution of several peptides in association with nerve fibers and ganglion cells mainly localized to the arterial vessels. Immunoreactivity was shown for neuron-specific enolase, neuropeptide Y, substance P, and vasoactive intestinal polypeptide [47]. Co-release of ATP may also occur based on studies of isolated blood vessels.

In vitro pharmacological studies of isolated hepatic blood vessels will not be extensively reviewed here. The portal veins are large, thin-walled vessels that offer virtually no resistance to flow except for the small intrahepatic tributaries. They have, inexplicably, been used as models of resistance vessels in some studies. In vivo plethysmographic recording of extrahepatic portal vein volume indicates a clear capacitance function with reflex activation being produced by cerebral ischemia [357]. This preparation may prove quite useful for studies of neuromodulators in vivo. The isolated large veins are responsive to nerve and drug-induced contractions, but there are quite remarkable species differences. The isolated rabbit portal vein, but not that of the guinea pig, shows ATP co-localized and released from sympathetic nerves. A nonadrenergic, noncholinergic inhibitory response, resistant to combined α- and β-adrenergic blockade and 6-hydroxydopamine-induced sympathectomy, is seen in the rabbit, but not the guinea pig, portal venous muscle [38,39]. In the isolated rat portal vein, adenosine serves as a prejunctional inhibitor of noradrenergic neurotransmission [36,167]. Adenosine is a potent dilator of the hepatic artery of the cat in vivo but does not affect portal pressure or portal responses to nerve stimulation [217]. The isolated rabbit hepatic artery also shows evidence of ATP cotransmission with noradrenaline [35]. The physiological relevance of the responses in these large conducting arteries is unclear. The frequency–response relationships in the isolated hepatic artery are grossly dissimilar from those seen in the cat, dog, and rat where the nerve frequency response that produces 50% of the maximal response is 1.7–3 Hz and maximal responses are reached between 6 and 20 Hz in all of these preparations. This is in contrast to the responses of the isolated hepatic artery where responses were rarely evoked at a stimulation frequency of less than 8 H, never below 4 Hz and frequency dependency was still seen up to 64 Hz.

11.10 DISEASE STATES

The contractility of isolated portal veins is enhanced in spontaneously hypertensive rats in response to noradrenaline and nerve stimulation. The intensity of response was increased at all doses and frequencies. The prejunctional modulatory effects of purines appeared unaltered as maximal responses were suppressed, but the EC_{50} for noradrenaline and Hz_{50} for nerve stimulation were not altered by an adenosine analogue, 2-chloroadenine [313].

Cirrhotic patients have elevated levels of circulating catecholamines of neural origin [132], and the levels correlate with the severity and prognosis of the disease and the extent of portal hypertension. The decrease in circulating noradrenaline and reduction in portal pressure with no change in hepatic blood flow after clonidine treatment [277,375] is compatible with the possibility that portal pressure elevation may be partly under sympathetic control in alcoholic cirrhosis. The adrenergic innervation appears grossly normal in diseased livers [178], but the normally dense catecholamine-specific fluorescence and AChE-positive terminals are reported to be absent after 4 days of bile duct ligation in the guinea pig; after 2 weeks, the nerve fibers could again be detected [365].

In the chronic bile duct ligated model of liver disease, hepatic compliance remained unaltered so that passive changes in liver volume in response to decreased perfusion pressure were similar in the control and diseased livers. In contrast, the response to hepatic sympathetic nerves appeared to be selectively decreased as the responses to norepinephrine remained unimpaired [335].

11.11 TECHNICAL CONSIDERATIONS (AVOID THE SUCKER PUNCH)

The considerations discussed in this section have been presented in several chapters and must be applied to any study of the peripheral vascular responses in the liver. These points are reiterated in this section because of the obvious technical artifacts that can be induced in studies of the effects of the hepatic sympathetic nerves and blood-borne constrictors.

Throughout this review, there is a strong emphasis on the fact that vascular reactions to passive influences may be as large as those that occur to active stimuli and that both active and passive influences occur simultaneously and interactively. Therefore, the technical procedures used to record vascular responses become extremely important. Each of the vascular responses has technical concerns that must be addressed. The hepatic artery, presinusoidal venous resistance, hepatic venous resistance, and hepatic capacitance responses are all subject to serious technical artifact.

Studies attempting to deal with regulatory processes acting on the hepatic artery often have the hepatic arterial buffer response as a complicating factor. To assess hepatic arterial tone in response to specific stimuli, the arterial perfusion pressure and portal venous flow must not change. An example of complex interactions and misinterpretation that can result from failure to control these parameters is shown by the following example [216]. When hepatic arterial blood pressure is held constant, reduction of portal venous flow, by occluding the superior mesenteric artery, results in a dilation of the hepatic artery secondary to the buffer response. At the same time, however, by decreasing portal flow using constriction of the superior mesenteric artery, systemic arterial blood pressure and, therefore, hepatic arterial blood pressure increase and the resultant increase in arterial flow washes away adenosine according to the mechanism of hepatic arterial autoregulation (see Chapter 5). When an adenosine receptor antagonist is used to block the buffer response, it also blocks the autoregulatory response so that constricting the superior mesenteric artery results in a similar increase in arterial pressure and a similar increase in hepatic arterial flow. From this example, it could be incorrectly concluded that blockade of adenosine receptors did not have a significant impact on the hepatic arterial buffer response [216].

The hepatic artery is affected by sympathetic nerves and by the hepatic arterial buffer response. Details of both responses have been previously discussed. It is important to note that in the in vivo preparation, direct electrical stimulation of the hepatic nerve plexus results in no significant blood flow change through the portal vein. Thus, the hepatic arterial response is a direct response to the nerve stimulation and is not complicated by a simultaneously activated buffer response. The use of vascular perfusion circuits to supply blood to the portal vein should be done in a physiological manner. An increase in venous resistance that does not lead to flow change in the venous circuit represents the in vivo situation. However, in many isolated liver perfusion preparations, the portal circuit is perfused under constant pressure, where active vasomotion leads to changes in portal flow. These changes in portal flow will be expected to produce a hepatic arterial buffer response, which will complicate interpretation of the direct effects of the stimulus on the hepatic artery. This type of problem is dramatically shown with responses to intravenously administered vasoactive compounds where the buffer response can overwhelm the direct effect of the drug and, in fact, produce the opposite change in vascular tone to what the drug produces on direct intra-arterial administration. The vasodilators glucagon, isoproterenol, and adenosine have been shown to elevate portal blood flow and reduce hepatic arterial flow at certain doses. When the portal blood flow change is prevented, only the direct dilator effect on the hepatic artery is revealed [208]. Similar complications would be anticipated to distort the response to nerve stimulation. The other major technical concern is that the changes in vascular tone of the hepatic artery should be estimated using calculated vascular conductance rather than resistance.

The portal venous resistance site is strongly influenced by sympathetic nerves but also the high distensibility of these resistance sites leads to them being dramatically influenced by the distending blood pressure. Changes in active tone can be determined using the calculated index of contractility as opposed to calculated vascular resistance. However, for many situations the net interaction between active and passive responses may be of importance and interest and the calculated resistance might be the most relevant measurement. In such cases, it is important to duplicate the most physiological condition so that the changes in distending pressure that occur in response to the active vascular tone adjustments will lead to appropriate passive effects on distension. For example, the effect on resistance could be dramatically different depending on whether blood flow is held constant or allowed to decrease. Nerve stimulation in vivo leads to marked elevation in intrahepatic pressure, whereas in a portal perfusion with pressure held steady, the reduced flow that ensues could be anticipated to result in increased heterogeneity of flow across the very low pressure gradient of the venous system.

Similarly, the interaction between stressed and unstressed volume within the liver and the net effect on capacitance must be carefully considered. Presumably, the effects on unstressed volume will be identical whether the blood flow or perfusion pressure is held steady. The consequences to stressed volume will, however, be quite different because the constant flow preparation will result in higher intrahepatic pressures. Reducing blood flow to the liver leads to large decreases in stressed volume.

Despite the fact that sympathetic nerve stimulation decreases hepatic arterial blood flow, it is not appropriate to attempt to assess the impact of this reduced arterial blood flow on the capacitance response using a mechanical occlusion of the blood flow to duplicate the flow pattern. Although the flow pattern would be similar in this situation, the intrahepatic pressure will decrease with the mechanical occlusion and result in a reduction of stressed volume, whereas in response to sympathetic nerve stimulation the intrahepatic pressure rises and stressed volume will be increased.

Another technical concern that must be addressed when attempting to measure stressed volume of the liver is related to the intrahepatic pressure measurements. Hepatic compliance is the measurement of the change in liver volume per unit change in the appropriate distending blood pressure. In earlier studies, we made the erroneous assumption that most of the resistance to blood flow resided in the portal veins and that the transmission of an elevation in central venous pressure to the compliant vasculature was virtually complete over a wide physiological range [212]. Subsequent studies have convinced us that the earlier interpretation was in error and that the primary resistance site is postsinusoidal and that the transmission of an elevation in central venous pressure is incomplete and dependent on the magnitude of the vascular resistance [223]. The lobar venous pressure measurement appears to be an acceptable estimate of intrahepatic distending pressure acting on the compliant capacitance vessels [121]. The use of portal venous pressure as an estimate of lobar venous pressure is accurate in basal situations where there is an insignificant pressure gradient between these two sites; however, it comes inaccurate under situations of active vasoconstriction where presinusoidal resistance can become significantly elevated and a pressure gradient develops between portal venous and lobar venous pressure.

In studying any homeostatic system, it is always imperative to be aware of the possibility of redundant control systems. A dramatic example of a redundant control system is the hyperglycemic response to hemorrhage induced by neuroendocrine redundant control systems in the liver. Hemorrhage results in a dramatic hyperglycemia of rapid onset that is not significantly affected by prior denervation of the liver. Without consideration of redundant control systems, one could conclude that the sympathetic nerves have no role in the control of the hyperglycemic response to hemorrhage. However, if the hepatic denervation is combined with bilateral adrenalectomy, the hyperglycemic response essentially disappears. In addition, if the adrenalectomy is carried out in the absence of hepatic denervation, the response is largely maintained. Thus, the existence of either the hepatic sympathetic nerves or the adrenal catecholamine secretions is capable of producing a normal hyperglycemic response to hemorrhage. A second redundant control issue has limited studies regarding the mechanism of hepatic blood volume changes in response to hemorrhage. Although it is known that hemorrhage results in activation of hepatic sympathetic nerves, secretion of adrenal catecholamines, generation of angiotensin, and elevated vasopressin levels, elimination of all of these active control systems results in an increased passive response of the liver to blood loss so that the compensation for blood volume removed is approximately 20% regardless of the presence or absence of these various active regulators. Whereas in the example of the hepatic arterial buffer response, technical manipulation of the required parameters is possible, this obstacle has not been overcome in regard to hepatic venous responses.

• • • •

CHAPTER 12

Hepatic Circulation and Toxicology

12.1 HEPATIC BLOOD FLOW

Hepatic clearance of many drugs is blood flow dependent. Compounds with high hepatic extraction show clearance rates that are directly dependent on hepatic blood flow. If fewer compounds are delivered to the liver, fewer compounds are removed from the circulation. In contrast, generally, compounds with lower extraction ratios tend to be blood flow independent [114,165]. Clearance of a number of important hormones, including aldosterone and corticosterone [265], is blood flow dependent. We have proposed that one of the major roles of the hepatic arterial buffer response is that this mechanism tends to maintain hepatic clearance of important endogenous compounds at a constant rate (Chapter 5). Normally, when considering endocrine regulation, one focuses on the rate of production and entry of the hormone into the bloodstream. Of equal importance, for the endocrine gland to afford fine control of hormone levels, a relatively constant background clearance rate of the hormone from blood must be maintained. If a hormone level is to be rapidly adjustable by altered output from the endocrine gland, rapid turnover of the hormone must occur. It is important that the high catabolic rate be maintained as constant as possible in order that the hormone levels can be accurately controlled by the glandular output. If hepatic blood flow was not prevented from rapid, transient changes secondary to similar changes in the portal venous flow, endocrine homeostasis would be imperiled. Thus, any toxic reaction that leads to an impaired hepatic arterial buffer response would be anticipated to result in subtle endocrine and metabolic disturbances.

Chronic alterations in hepatic blood flow occur with the normal aging process. Clearance of a large number of drugs declines with age. The decline does not appear to be attributable to reduced activity of drug metabolizing enzymes, with the possible exception of a few highly specific cytochrome P450 forms. Hepatic blood flow declines significantly with aging and is suggested to account for the predominance of reduced hepatic drug clearance in the elderly [336,380].

Clearance of drugs with a marked first-pass effect is decreased in the presence of a portacaval shunt. The bioavailability of many compounds, including β-adrenergic antagonists, calcium channel blockers, several analgesics, and anxiolytics, will be increased in patients with a surgical portacaval shunt or endogenous portal systemic shunts, such as seen in cirrhosis. Doses of such compounds are recommended to be adjusted downward to avoid toxicity [29,126]. Creation of a portacaval shunt leads to hepatic atrophy and reduced levels of cytochrome P450. A number of liver enzymes are not affected; the activities of four mixed function oxidases are reduced and the heme oxygenase rate-limiting enzyme in catabolism of heme to bilirubin is enhanced [78]. Considerable differences in drug metabolism and, possibly, toxicity would be anticipated in patients with portacaval shunts or reduced hepatic flow due to age.

The toxic effect of drugs on the liver has often been linked to disturbance of blood flow. The phenothiozine derivatives, thioridizine and chlorpromazine, increase hepatic vascular resistance in cats, rats, and dogs, leading to elevations in portal pressure, decreases in cardiac output, and portal flow [303]. A single clinically relevant dose of the immunosuppressant cyclosporine led to a greater decrease in total hepatic blood flow than what is seen for renal blood flow where the renal vasoconstriction has been attributed with at least partially accounting for nephrotoxicity. The potential impact on hepatic function has not been fully evaluated [9].

Much of the postsurgical hepatic complications seen in patients may be related to hepatic circulatory disruption caused by the anesthetics. Gelman [92] reviewed the area of general anesthesia and hepatic circulation. He concluded that most of the anesthetics decrease portal blood flow in association with a decrease in cardiac output but that hepatic arterial blood flow can be preserved, decreased, or increased. Hepatic oxygen deprivation is believed to be a significant factor in the production of hepatotoxic reactions to anesthetics. Although it has been suggested that anesthetic agents be ranked by their ability to preserve hepatic blood flow in relation to hepatic oxygen consumption, no detailed formalized ranking appears to exist. Halothane appears to be the most hepatotoxic compound and the toxicity may relate to the dramatic hepatic circulatory disruptions seen, with decreases occurring in both portal and hepatic arterial flow and a reduced hepatic arterial autoregulation and hepatic arterial buffer response [7,8,341].

Toxicant interactions with blood flow may be related primarily to the oxygen delivery and alterations in free radical formation that are oxygen-dependent (discussed later). The role of circulatory dysfunction within the liver for potentiation of hepatotoxic effects can be assumed from studies demonstrating the importance of oxygen supply to toxicity. Shingu et al. [343] demonstrated that rats pretreated with phenobarbital showed potentiated hepatic injury in response to a variety of anesthetics including halothane, enflurane, isoflurane, thiopental, and fentanyl when administered 10% oxygen. No agent given with 20% or 100% oxygen demonstrated hepatotoxicity. Liver temperature was also demonstrated to be important in determining hepatotoxicity, with mild hypothermia (32–34°C) preventing hepatotoxicity by enflurane and isoflurane but not by halothane when administered with 10% oxygen [343]. The hepatic arterial buffer response has been shown to be maintained in livers with severe disease [11,127,152,282,316,330]. Although the HABR is seen in cirrhotic livers, the buffer capacity may be insufficient to maintain a normal oxygen supply [332].

Ischemia or hypoxia may lead to temporary protection of hepatocytes in some instances. Cellular acidosis, a consequence of ischemia or hypoxia, is protective against death of hepatocytes from a variety of toxicants, including oxidant chemicals, ionophores, and mitochondrial inhibitors [286]. During injury, hydrolytic enzymes with neutral or alkaline pH are activated. Their activities are strongly inhibited by acidotic pH, but reperfusion, which causes a rise of intracellular pH, releases this inhibition leading to reperfusion injury [263]. Adenosine, which is produced by an ischemic or hypoxic liver, is an effective hepatoprotector against the liver damage induced by CCl_4, ethanol, and cyclohexamide [89,133]. These agents cause an increase in oxidant stress. In the presence of adenosine, free radical-induced damage was decreased, but it is suggested that the protective effect is not directly due to adenosine but rather to free radical scavenging catabolic products of adenosine, such as uric acid [50]. It is unlikely that adenosine's protective effect is mediated by an increase in hepatic blood flow because adenosine produced by hepatic parenchymal cells is downstream from the arterial resistance vessels and will not, therefore, have access to the resistance vessels. This is consistent with the anatomy of the acinus and the observation that reduced pO_2 in hepatic arterial perfusate leads to the hepatic release of adenosine without producing significant arterial vasodilation (Lautt and Legare, unpublished observations).

12.1.1 Veno-Occlusive Toxins

A number of toxins, seen in plants consumed as herbal remedies or teas, or that appear as contaminants in foods, produce severe occlusion of the hepatic veins leading to hemorrhagic necrosis mainly restricted to zone 3 of the acinus. Serious outbreaks of veno-occlusive disease have been reported in Jamaica and India where contamination occurred from seeds of plants of the senecio species contaminating commercial grain stocks [358]. In Jamaica, heavy use of "bush teas" especially comfrey, used as a sedative for infants, has resulted in many severe cases. It has been suggested that compounds such as azathioprine, monocrotaline, and dacarbazine produce hepatic veno-occlusive disease primarily as a result of injury to the endothelial cells subsequent to a profound depletion of glutathione. Supplemental glutathione protects against these toxicants. It is possible that glutathione exported from hepatocytes may serve a protective function for endothelial cells that would be exposed to the glutathione concentration in the space of Disse before the glutathione is diluted in the sinusoidal circulation. Reduced glutathione concentration in hepatocytes might be anticipated to potentiate toxicity of compounds that act primarily on endothelial cells [67].

12.1.2 Heterogeneity of Perfusion

Heterogeneity of blood flow occurs at different levels within the liver. The most obvious heterogeneity occurs at the smallest vascular unit, the sinusoid. One form of heterogeneity that is seen in this situation is related to the unique microvasculature of the hepatic acinus whereby mixed portal and hepatic arterial blood enter the sinusoids in zone 1 at the center and flow outward to drain from zone 3. Autonomic innervation of hepatocytes is most dense in zone 1, as is the population of Kupffer cells. The heterogeneity in this case is primarily related to the constituents of the blood and the changes that occur in the constituents as the blood flows through the sinusoids, having compounds removed or added. The trans-acinus gradients can be extremely steep, positively and negatively. The gradient of oxygen can result in regulation of enzyme selectivity in different zones.

Liver damage induced by many xenobiotics is zone-dependent (reviewed by Jungermann and Katz [157]) with perivenous (Rappaport's zone 3) necrosis being noted for bromobenzine, ethanol, and CCl_4. Periportal (zone 1) lesions are seen, for example, with allyl alcohol and digitonin. Retrograde perfusion of isolated livers results in the lesions from these latter compounds appearing in the perivenous zone, thus potentially implicating oxygen tension as an important factor. Drug metabolism is differentially regulated in zones 1 and 3, and chronic changes in blood flow are known to alter enzyme profiles. Perfusing the liver with heavily oxygenated blood, with a predominantly arterial supply, results in zone 1 cell types encroaching into the central zone with zone 3 cell characteristics decreasing as a proportion of the cell population. Similarly, a reduction in chronic oxygen or arterial blood flow perfusion leads to the reversed situation with zone 3 population expanding and zone 1 characteristics diminishing in proportion [309].

Heterogeneity, as the term is usually used, is also seen within the liver, again most obviously at the smallest unit. Individual sinusoids show episodic flow varying from swift passage of red blood cells to the opposite extreme of temporarily fully stagnant sinusoids. Temporary reversal of sinusoidal flow or even circular flow around a few parenchymal cells has also been demonstrated.

The capsule of the liver receives approximately 25% more arterial flow than deeper tissues [232]. Portal flow distribution within liver lobes most often shows a pattern of highest flow to the top, graduating down to lowest flow to the bottom (the dorsal side on animals placed on a surgical table). This degree of heterogeneity of perfusion appears to be under dynamic and multiple interacting forces [232]. The significant heterogeneity shown for flow on immediately adjacent portions of the liver surface indicates that extreme caution is needed when using small areas of flow determination to reflect changes to the entire organ. The functional implications of the gross levels of heterogeneity are not clear. However, heterogeneity does not appear to be significantly increased in studies of large animals under in vivo conditions even under conditions of low flow, norepinephrine infusion, sympathetic nerve stimulation, or raised venous pressure [59,119,120].

It is feasible that microcirculatory heterogeneity that does not alter total flow could have toxicological significance based on the role of oxygen in free radical formation. Carbon tetrachloride-induced damage is associated with an increased sympathetic nerve activity that results in small changes in hepatic blood flow. In the spontaneously hypertensive rat, CCl_4 resulted in a much larger activation of sympathetic activity and a decreased hepatic blood flow concomitant with significantly enhanced liver damage. It was suggested that regional ischemia may account in part for some of the toxic reaction to CCl_4 [140]. In normal cats, acute administration of CCl_4 does not result in significant changes in hepatic arterial or portal venous flow over the first 4 h when liver damage is clearly seen. Similarly, at 24 h, there was no evidence for altered total hepatic flows [230]. Regional heterogeneity of flow leading to local ischemia or hypoxia may potentiate CCl_4-induced damage but is clearly not a primary requirement and does not reflect as altered total flow.

Although it is unclear if the effects are due to circulatory disruption, it is clear that activation of sympathetic nerves during exposure to toxicants can potentiate liver damage [150]. In addition, prior exposure to toxicants leads to a state of liver physiology that responds to strong sympathetic nerve stimulation or norepinephrine infusion by augmenting liver cell damage. Although the effect of autonomic nervous activity in the liver and the involvement of the hepatic circulation have not been evaluated from a toxicological perspective, certain lines of evidence indicate that the hepatic sympathetic nerves and circulating catecholamines can have considerable impact both with exacerbating acute liver damage and with potentiating damage that has previously occurred. Foot shock stress increased sympathetic activity in the liver and accelerated chemically induced liver injury in rats [149]. Electrical stimulation of the ventral medial hypothalamus also activates hepatic sympathetic nerves and potentiates acute liver damage induced by CCl_4 [151]. Enzyme markers of acute liver damage were increased during direct electrical stimulation in isolated livers or during noradrenaline perfusion in livers pretreated 24 h previously with galactosamine. Although hepatic sympathetic nerves have been shown in isolated liver preparations perfused with nonblood perfusate to release prostaglandins and cytokines from nonparenchymal cells, prostaglandins do not seem to be involved in the potentiating effect of nerve stimulation on acute liver injury [150]. The mechanism of nerve-induced potentiation of liver damage is not known, but the effect is dependent on extracellular calcium. The effect of autonomic activation also may be different for acute toxicity compared to chronic toxicity because it has been reported that exogenous norepinephrine provided protection against chronic CCl_4-induced liver damage by an unknown mechanism [144].

12.1.3 Capillarization

A further hepatic vascular dysfunction occurs through the process known as capillarization where chronic exposure to toxic compounds leads to a decrease in fenestration size, with the endothelial cells becoming similar to capillaries of other organs. Kupffer cell concentration is reduced in capillarized sinusoids [246].

The first drug shown to alter fenestrae size was ethanol, which initially dilates the fenestrae and may partially account for the protective effect from atherosclerosis ascribed to low-dose ethanol consumption. Long-term alcohol abuse, however, leads to capillarization of the sinusoidal endothelium. Capillarization is a reversible process so that 4 months after discontinuation of thioacetamide, the sinusoidal fenestrations were increased and, by an additional 6 months, the basement membrane in the space of Disse disappeared in rats [278].

It has been suggested that the decrease in the number and size of fenestrae in the endothelial cells, and the appearance of basement membranes in the space of Disse that is seen in alcoholic liver disease, may cause a disturbance in exchange of many bioactive substances between the sinusoidal blood and hepatocytes across the Disse space and may thereby contribute to the pathogenesis of alcoholic liver disease [366]. An alternative hypothesis has been presented by Hickey et al. [134] who proposed that the impaired hepatic function attributed to capillarization is not secondary to impaired access of the drug to the hepatocyte but rather to impaired oxygen delivery to the hepatocyte. This proposal is supported by the observation that the metabolism of oxidized drugs is impaired in cirrhosis, whereas drug glucuronidation is relatively spared. In the healthy liver, oxidative drug metabolism is more sensitive to hypoxic conditions than glucuronidation [10]. Theophylline clearance was reduced in cirrhotic rats, and this reduction could be reversed to levels seen in healthy animals by oxygen supplementation in the respired air. Oxygen supplementation did not result in increased theophylline clearance in normal animals [134].

Although the effect of capillarization on the ability of drugs to diffuse to the hepatocyte may be controversial, clearly the alteration of fenestrae numbers and size has the potential to cause serious metabolic disturbances with lipid metabolism. The fenestrated hepatic sinusoidal endothelium serves as a sieve that separates the hepatic blood from the plasma in the space of Disse. The fenestrations are of such a size that they act as a barrier to the large triglyceride-rich chylomicrons and, only after the chylomicron has diminished in size, secondary to triglyceride uptake in adipose tissue, is the remnant chylomicron sufficiently small to pass through the fenestrations to contact the hepatocyte (reviewed by Fraser et al. [86]). The triglyceride exported from the liver as very low-density lipoprotein in the hepatic lymphatics might be maintained, whereas the hepatic uptake of circulating cholesterol (low-density lipoprotein) could be severely compromised. This liver sieve serves as a site for regulation of hepatic selection and metabolism of dietary cholesterol and fat-soluble vitamins. The ability of the hepatic sieve to partially restrict albumin access to hepatocytes in healthy livers [101] implies that capillarization could lead to reduced availability of protein-bound compounds to hepatocytes. The disturbance of lipid and endocrine homeostasis may contribute significantly to liver disease and some of the dysfunctions in other organs that are seen in liver disease [86].

12.1.4 Ethanol

The relationship between acute and chronic effects of alcohol and the hepatic circulation is not clear. In reviewing the mechanism of ethanol-induced hepatic injury, Lieber [240] took issue with the hypoxia hypothesis that proposes that alcohol leads to an increased consumption of oxygen that results in a reduction of oxygen tension along the sinusoids to the extent of producing anoxic injury in zone 3 of the acinus. In the literature reviewed, he indicates that the studies showing decreased hepatic venous oxygen saturation and tissue oxygen tensions were within normal ranges and that, in a number of studies, increased oxygen uptake was accompanied by increased portal venous flow (secondary to increased flow of extrahepatic splanchnic origin) such that zone 3 hypoxia was very unlikely. Acute ethanol was reported to result in a significant increase in hepatic blood flow and an increase in oxygen consumption but with the net result being increased hepatic venous oxygen content [369].

Many of the toxic effects of ethanol may be linked to the ability of ethanol to shift the redox equilibrium toward a reduced state. The dependence of the ethanol-induced redox shift on oxygen tensions may contribute to the selective zone 3 hepatotoxicity of alcohol. Anesthetized cats showed no stimulation of ethanol metabolism or enhanced oxygen uptake after acute or chronic administration of ethanol [115]. No evidence for a hypermetabolic state induced by chronic ethanol administration was seen in innervated or acutely denervated livers. Oxidation of ethanol was estimated to require 40–45% of normal oxygen uptake, leading to the conclusion that other oxidative processes must be suppressed during ethanol metabolism. Hepatic lactate uptake remained unaltered when ethanol metabolism was less than 0.5 of V_{max} but was suppressed on an equimolar basis with the increase in ethanol metabolism when ethanol metabolism rose to greater than 0.5 V_{max} [115].

Despite the large number of studies, beyond the scope of this chapter, the blood flow responses to acute alcohol exposure vary from decreased to increased; oxygen uptake has been reported to decrease, increase, show no change, or have biphasic effects [347]. Some of the differences are clearly methods-dependent as shown by increased portal flow in conscious but not anesthetized rats [48]. Dose and species and protocol differences also exist. Much of the damage imposed by ethanol appears to be related to free radical formation because a variety of antioxidants reduce or completely prevent the ethanol-induced damage. These compounds include vitamin E and *S*-adenosyl methionine [5]. The involvement of free radicals in toxic reactions is discussed later.

12.1.5 Methodological Considerations

For technical reasons, many drug metabolic studies and acute toxicity studies were carried out using isolated perfused liver preparations. A substantial body of evidence indicates that oxygen content of the perfusate can alter the toxic reaction to drugs and that ischemia results in activation of a number of autocoids released from within the liver that can have very dramatic actions on the hepatic vasculature and metabolic functions. Most isolated perfusion studies do not use a perfusate containing red blood cells, so that the pO_2 of the blood reaching zone 1 is excessively high, whereas the pO_2 of blood exiting zone 3 is abnormally low compared with in vivo levels. Similarly, most studies do not perfuse through the hepatic artery, thus eliminating a normal regulatory mechanism designed to produce homogeneous distribution of blood throughout the hepatic microcirculation.

Additional concerns related to studies of the interactions of metabolism, toxicity, and blood flow are based on the fact that vasoconstriction within the liver normally will result in an increase in portal pressure but portal flow will not decrease [213]. If the vasoconstriction occurs at postsinusoidal sites, the consequence will be an elevated intrahepatic pressure with sinusoids being prevented from passive collapse. In contrast, in most perfusion studies, an elevation of intrahepatic vascular resistance occurs during constant pressure perfusion, with the result being that vasoconstriction results in a decrease in hepatic blood flow. Intrahepatic pressure will not rise, and it is anticipated that the decreased flow within this extremely low pressure circuit would result in increased heterogeneity of perfusion. Pharmacokinetic studies carried out in such preparations must be interpreted with caution. For example, increasing hepatic extraction seen with increased blood flow could merely represent increased sinusoidal perfusion of sinusoids previously underperfused.

The primary limitation of in vivo studies is related to the need to use anesthetics. Anesthetics, such as halothane, interfere with normal hepatic vascular regulation [7,8,92,341]. Increased portal flow is seen in response to oral ethanol in conscious but not anesthetized animals [48]. Although some anesthetics are clearly hepatotoxic, such as halothane, other anesthetics appear to have less interactions, but all of them clearly produce some abnormal functioning [92]. If anesthesia is required, sodium pentobarbital is well suited for homeostatic neurovascular and metabolic studies.

Stress results in alterations of hepatic circulation and hormonal milieu, which may have considerable significance in toxicological studies. It has been demonstrated that activation of sympathetic nerves leads to potentiation of toxicant-induced liver damage discussed earlier. It is known that surgery results in significant disturbance of gastrointestinal tract motility and blood flow, which may fully return to normal only after several days [294]. The common use of short-term anesthesia to instrument an animal, followed by studies carried out within several hours of recovery from the anesthesia, incorporates most of the disadvantages of both anesthetized and unanesthetized preparations. These studies should be carefully evaluated for comparison with results obtained after several days of recovery.

12.2 FREE RADICALS AND ANTIOXIDANTS

From previous sections it becomes clear that regional ischemia leading to hypoxia can potentiate the hepatotoxic effects of a variety of compounds. Similarly, a number of stimuli that are reported to increase microvascular perfusion heterogeneity, such as stress and sympathetic nerve activation, are capable of potentiating hepatotoxins. Kupffer cell activation, either by drugs or by prior administration of endotoxin, leads to microvascular disturbances, and prevention of such disturbances offers protection against a variety of toxicants [34,65,242]. Although these various observations cannot uniformly be directly linked to free radical formation, many can. Several reviews can be consulted for specific references [164,295,388].

A broad definition of a radical is that it is a molecule or ion containing an unpaired electron. Most radicals are extremely reactive and undergo rapid reaction in which the unpaired electrons become paired. Some radicals however are relatively stable and have long half-lives. The reactivity of a radical lies in the speed and affinity with which the unpaired electron participates in covalent bonding. Often, the interaction of one radical leads to the generation of a series of other steps of radical formation before final pairing. Activation of oxygen for participation in metabolic pathways results in the reactive radical intermediates, the superoxide anion radical, hydrogen peroxide, and the hydroxyl radical.

A number of enzymatic systems have evolved that can scavenge or eliminate excess radicals from metabolically active biological tissue. Most oxidative metabolism occurs within mitochondria, which also serves as the major intracellular source of superoxide radicals. Superoxide dismutases reduce superoxide radical to hydrogen peroxide, which is further decomposed to water by catalase and a variety of peroxidases including glutathione peroxidase. These enzymes serve as a scavenging system in which each enzyme plays an integral role in free radical modulation. The free radical scavenging systems are numerous and interactive.

Lipid peroxidation reactions can be initiated by radical species, resulting in chain-propagating lipid radical reactions that can release lipid hydroperoxides and significantly alter membrane fluidity and functions [91]. A variety of hydrophobic scavengers such as vitamin E and β-carotenes are incorporated into cellular membranes and inhibit chain-propagating reactions in lipid microenvironments [41]. A number of glutathione-dependent scavenging systems are capable of reducing lipid hydroperoxides; however, they appear to require involvement of the hydrophobic scavengers to transfer the reagents from the lipid membrane to the glutathione activity, which is represented by soluble enzymes.

Antioxidant mechanisms also operate effectively in the hydrophyllic tissue environment. The oxidation of extracellular, cellular, and membrane-bound proteins can be caused by a number of the strong oxidants generated by radical reactions. Molecules, such as ascorbic acid, cysteine, and reduced glutathione, all play a role in the prevention of this oxidation and the regeneration of normal protein structure.

A number of oxidants or oxidant potentiating compounds and antioxidants are unwittingly ingested with normal and industrial society-modified diets, and the relative dietary intakes of these compounds may well account for much of the individual variability in toxicant impact. Diets that are extremely high in polyunsaturated fatty acids (as a result of inappropriate attempts to modify the levels of saturated fatty acids in the diet) have led to increased concentrations of the polyunsaturates in cell membranes and increased oxidant products of these fatty acids. Elevated levels of peroxidized polyunsaturated fatty acids can be readily prevented by administration of vitamin E or vitamin C. High polyunsaturated fatty acid content in the diet has been associated with potentiation of ethanol-induced damage and cell death produced by carbon tetrachloride and other toxins [68]. Lipid peroxidation has also been attributed with the changes in surface receptors for a variety of hormones in the liver and peripheral tissues including insulin, glucagon, and epidermal growth factor [21]. Treatment with a variety of antioxidants, including cinnamon extracts [281], has a hepatoprotective effect against ethanol and carbon tetrachloride administration [70]. Patients with chronic liver disease have a high incidence of systemic endotoxemia that may be anticipated to result in oxygen radical formation [242] and contribute to general systemic deterioration.

The relationship of hepatic anoxia or ischemia to oxygen-derived free radicals has been clearly demonstrated in ischemia-reperfusion injury [259,296,297]. Although the duration of time required to show this sort of ischemia-reperfusion injury is not well delineated, and it is unclear whether such effects would occur as a result of regional ischemia in small zones of the liver, overall antioxidant status and oxidant load will clearly impact on the tissue damage induced by oxidant producing toxicants.

S-adenosyl-L-methionine (SAMe) is a metabolite in the trans-sulfuration pathway for the metabolism of methionine. This pathway, chiefly in the liver, uses more than 70% of dietary methionine and is responsible for the synthesis of polyamines, cysteine, glutathione, and taurine, as well as a large number of methylated molecules [45,352]. Among the methyl group acceptors are proteins, phospholipids, DNA, RNA, histones, a number of hormones, and creatine. In the process of methylation, SAMe is converted to *S*-adenosyl-homocysteine, which is further degraded to adenosine. Note that our current hypothesis for the origin of adenosine involved in the regulation of the hepatic arterial buffer response is through this pathway. SAMe has undergone extensive clinical trials for a surprising range of clinical conditions including the treatment of neuropsychiatric diseases, degenerative diseases, cholestasis, and chronic liver disease of a variety of etiologies. The production of SAMe in cirrhosis is reduced to approximately 30% of normal [44]. Exogenous SAMe appears to not only correct the plasma deficiencies of choline and cysteine but may also correct other hepatic methylation processes [51]. Parenterally administered SAMe antagonizes liver damage in rodents produced by D-galactosamine [353] or paracetamol [32] and prevented damage or accelerated recovery from steatosis produced by choline deficiency or alcohol. Hexachlorobenzine is a worldwide environmental pollutant known to be a potent hepatotoxic compound that causes a type of porphyria characterized by accumulation of phorphyrins with a high number of carboxylic groups in the liver. Hexachlorobenzine has been shown to reduce stores of the potent antioxidant, glutathione; and administration of SAMe offers significant protection against hexachlorobenzine [46]. Whether the protective effect is related to restoration of glutathione stores or to restoration of methylation or to the combination of antioxidant effects is not clear.

Other potent antioxidants have been demonstrated in a number of models to protect against a variety of hepatic insults that appear to be secondary to radical formation. Primary among these is vitamin E, which has been shown to be protective against carbon tetrachloride [298]. β-Carotene is reported to have a considerably improved antioxidant capacity in conditions of low oxygen partial pressure including ischemia [169]. Although ischemia is shown to potentiate toxic effects of agents believed to produce hepatic dysfunction secondary to radical formation, a direct contributory role of total or regional ischemia in toxic reactions to xenobiotics and protective or restorative effects of increased arterial blood supply remains unevaluated.

12.2.1 Samec

The generally unimpressive performance of antioxidants in clinical trials stands in dramatic contrast to the massive amount of epidemiological evidence indicating that antioxidants consumed in food confer demonstrable positive effects in multiple pathologies. Antioxidant clinical trials invariably involve administration of only one or two antioxidants. While carrying out studies related to the hepatorenal syndrome (Chapter 13), we developed a research tool to attenuate the acute (24-h) toxicity of thioacetamide, well recognized to cause hepatotoxicity through the generation of free radicals. Liver damage was assessed from histological appearance and hepatic enzyme levels and by interference with meal-induced insulin sensitization. Intraperitoneal thioacetamide resulted, 24 h later, in hepatic inflammation, elevated hepatic enzymes, and abolition of meal-induced insulin sensitization. The decrease in postprandial responses to pulses of insulin acting on skeletal muscle is a well-described pathway involving feeding signals from the gastrointestinal tract acting on the liver through hepatic parasympathetic nerve activation and a second signal, an elevation of approximately 40% in hepatic glutathione levels. In the presence of the parasympathetic nerve and glutathione signals, a pulse of insulin stimulates the release of a hepatic insulin-sensitizing substance (HISS) from the liver. Both of these "feeding signals" are necessary but neither is sufficient alone to allow a pulse of insulin to cause the release of a pulse of HISS from the liver. HISS acts selectively on skeletal muscle. A full description of the HISS hypothesis is beyond the scope of this chapter, but recent reviews provide a full referenced background including the hypothesis that absence of HISS action after a meal is the first metabolic defect leading progressively to obesity, syndrome X, and type 2 diabetes (reviewed in references [200–203,228]). Provision of the combination of vitamins C and E 1 h after the administration of thioacetamide did not provide significant protection nor did the administration of SAMe when given alone. However, the combination of the antioxidants resulted in very significant protection according to all indices. The rationale for the specific combination was based on the use of SAMe to protect the hepatic glutathione levels, vitamin E to protect the lipid constituents of the cell, and vitamin C to act primarily in the aqueous phase. With these three major sites targeted for protection, dependent on the specific characteristics of the antioxidant, free radical quenching appeared to be synergistically enhanced by the simultaneous targeting of multiple cellular sites of radical dysfunction. This synergistic antioxidant cocktail, referred to for convenience as Samec, has also been shown to provide chronic protection against the development of liver-dependent peripheral insulin resistance that is seen with aging [229] and a diet supplemented with sucrose [273]. It is predicted that absence of meal-induced insulin sensitization results in a shift of the homeostatic balance of nutrient partitioning away from glycogen in skeletal muscle and toward fat. In the absence of HISS, the response to a pulse of insulin is reduced by more than 50% in rats and humans. The result of absence of meal-induced insulin sensitization is that every meal then results in postprandial hyperglycemia, which is reacted to by hyperinsulinemia. Insulin is a lipogenic hormone acting on adipose tissue and liver to form triglycerides and lead to adiposity. The mechanism of nonalcoholic fatty liver disease is highly likely to be explained by absence of meal-induced insulin sensitization. Although absence of meal-induced insulin sensitization has been directly linked to adiposity and other dysfunctions associated with the metabolic syndrome and type 2 diabetes, the effect on hepatic circulation has not been specifically examined. The effect of nonalcoholic fatty liver disease on hepatic vascular homeostasis has not been examined, but the use of the antioxidant cocktail, Samec, may be a useful research tool to determine both the impact on fatty liver and the impact of fatty liver disease on other homeostatic systems in the liver including the microvasculature.

Although Samec was designed as an antioxidant cocktail to specifically protect vulnerable regions of the cell, it is not clearly demonstrated that the effects are mediated through free radical management. As discussed elsewhere in this monograph, SAMe is involved with a great many chemical reactions and homeostatic systems within the liver and elsewhere.

• • • •

CHAPTER 13

Hepatorenal Syndrome

Patients who die of liver disease die in renal failure. Renal dysfunction is demonstrable at the early stages of liver disease. As liver injury progresses, functional renal failure develops, resulting in sodium and water retention, and decreased renal blood flow and glomerular filtration rate, in the absence of significant morphological changes in the kidney. Various mechanisms have been suggested for the pathogenesis of renal insufficiency secondary to acute and chronic liver injury including peripheral arterial vasodilation secondary to overproduction of vasodilator substances in the splanchnic circulation, leading to splanchnic pooling and decreased effective systemic arterial plasma volume [76,93,94,95]; overproduction of endothelin due to endotoxemia leading to renal vasoconstriction [76,93,94,95]; and activation of a hepatorenal baroreflex that stimulates renal sympathetic nerves, leading to sodium retention [69,94,153,171,182]. We have recently suggested that a hepatic blood flow-dependent hepatorenal reflex is the primary pathophysiological mechanism for renal dysfunction in liver disease. This reflex is activated by adenosine, in the space of Mall, which is regulated by hepatic blood flow [269,270,274,275].

It has long been recognized that in hepatic cirrhosis, the disturbance in hepatic portal circulation relates to the pathogenesis of sodium and water retention through the activation of a hepatorenal reflex [190]. Liver cirrhosis is characterized by increases in renal sympathetic nerve activity [69]. Selective bilateral renal denervation, produced by lumbar sympathetic anesthetic block, promotes renal water and sodium excretion in these patients [346]. Animal models of cirrhosis show an increase in renal efferent sympathetic nerve activity that contributes significantly to the pathophysiological renal retention of sodium and water resulting from activation of a hepatic afferent limb [1,168,322]. Although the efferent limb of the renal disturbance is reasonably defined, the afferent limb has, until recently, remained unclear.

A consensus appears to have arisen that the intrahepatic vascular resistance that occurs in chronic liver disease results in portal hypertension with the elevated portal pressure serving as the afferent limb of the hepatorenal reflex. However, such a reflex implies a positive feedback situation, whereby an increase in portal blood flow would cause an increase in portal pressure and activation of the hepatorenal reflex. This would result in salt and water retention and an expanded blood volume, leading to increased cardiac output and increased portal flow, with a further increase in portal pressure. Such a positive feedback would serve no useful homeostatic function. The alternate hypothesis, that portal flow is the sensed parameter regulating the hepatorenal reflex, had not been previously suggested. In fact, there had never been a suggestion of regional blood flow being monitored by sensory nerves in any organ. As with any paradigm change, anomalies in the old paradigm had increasingly appeared. A number of earlier studies had suggested that the hepatorenal reflex was unlikely to be activated in response to baroreceptors. Using anesthetized dogs, Koyoma et al. [171] observed that the partial occlusion of the portal vein resulted in activation of renal sympathetic nerves that was not related to increases in either extrahepatic portal pressure or intrahepatic sinusoidal pressure (because intrahepatic sinusoidal pressure was decreased in these studies). Levy and Wexler [238] found that sodium retention persisted in cirrhotic dogs after end-to-side portacaval anastomoses, a maneuver that normalized intrahepatic hypertension but was still associated with a dramatic decrease in intrahepatic portal blood flow. Liang [239] reported a lack of correlation of increased portal pressure with the rate of urine flow at portal pressure elevations up to 15 cm H_2O; only at pressures above this level, when portal blood flow would have been reduced, did the urine flow rate begin to decrease. Most of the studies purporting to show evidence for portal pressure regulation of the hepatorenal reflex have also resulted in reduction of intrahepatic portal flow. Cirrhosis is characterized by a hyperdynamic splanchnic circulation and portal hypertension [27,94] but, because of the presence of portacaval shunts directing flow around the liver, the blood flow that directly perfuses functional sinusoidal and parenchymal hepatocytes is actually decreased [163].

The hypothesis relating intrahepatic blood flow to the hepatorenal reflex is supported by a recent series of publications and ongoing studies reported by us. The hypothesis is that reduced functional portal blood flow through the liver results in reduced washout of adenosine from the space of Mall (as described in Chapter 5 related to the hepatic arterial buffer response). Adenosine acts on sensory nerves arising in the space of Mall and activates the hepatorenal reflex. The hypothesis was tested progressively.

We established a vascular shunt connecting the portal vein and vena cava in rats to allow for control of the portal venous blood flow [275]. Partial occlusion of the portal vein, close to the hilum of the liver, decreased intrahepatic portal flow and the extra portal flow was allowed to bypass the liver through the shunt to prevent splanchnic congestion. A 50% decrease in intrahepatic portal flow through this mechanism did not cause significant changes in systemic arterial blood pressure but decreased urine flow by 38% and sodium excretion by 44%. The renal effect of reduced portal blood flow was prevented by hepatic denervation or intraportal administration of the adenosine receptor antagonist, 8-phenyltheophylline. Involvement of intrahepatic baroreceptors was eliminated because intrahepatic sinusoidal pressure was decreased after partial portal vein occlusion. These studies provided the first evidence that intrahepatic portal flow could activate a hepatorenal reflex.

Our prior studies related to the HABR indicated that adenosine in the space of Mall was regulated by intrahepatic blood flow. The observation that the hepatic perivascular region is also rich in sensory nerves [288] supported the feasibility of an adenosine-mediated afferent limb in the hepatorenal reflex. Adenosine has previously been shown to activate sensory nerves in the carotid body [367] and in the heart [359]. Stimulation of myocardial adenosine A_1 receptors increased the discharge of cardiac afferent fibers and resulted in an increase in neural discharge of the renal sympathetic efferent fibers in anesthetized dogs [276,359]. To test if adenosine could activate a hepatic afferent reflex, adenosine was infused directly into the portal vein and resulted in a significant decrease in urine flow and sodium excretion. In contrast, intravenous adenosine at the same dose was without any effect on renal function, thereby indicating that the effect of the infused adenosine was through the liver and not a direct action on the kidney. Intraportal infusion of the adenosine receptor antagonist, 8-phenyltheophylline, abolished the renal response to intraportal adenosine. Furthermore, both hepatic and renal denervation abolished the renal response to adenosine, thereby proving the reflex connection (as opposed to a possible hormonal connection) [274]. Thus, these data taken together are consistent with the hypothesis that reduction in intraportal blood flow leads to an adenosine-mediated activation of hepatic afferent nerves, which results in a sympathetic reflex to the kidneys, leading to fluid retention.

This response would serve a useful function in normal physiological conditions where the reduced portal flow would cause fluid retention, thereby increasing the circulating blood volume and cardiac output. The elevated cardiac output would result in elevated portal flow, thus correcting the flow imbalance to the liver. The hypothesis also proposes that, in the diseased state, with portacaval shunts existing, the signal would be anticipated to occur as a result of the decreased intrahepatic portal flow. However, in this state, the salt and water retention would not lead to a correction of the intrahepatic flow but, rather, would lead to elevated cardiac output and elevated portal inflow (the hyperdynamic circulation), which would simply bypass the liver through the shunts and lead to a progressive, inappropriate reflex accumulation of fluid.

We have recently demonstrated that renal dysfunction is mediated through this adenosine-dependent hepatorenal reflex in both acute and chronic liver disease models in rats. Chronic administration of the hepatotoxin, thioacetamide, resulted in severe fibrosis consistent with advanced liver disease (Chapter 12, hepatic circulation and toxicology). Reduced basal urine flow and a reduced ability to excrete a saline load were demonstrated. The renal dysfunction was partially corrected by intrahepatic administration of the adenosine receptor antagonist, 8-phenyltheophylline [269]. An acute model of liver injury involved intraperitoneal injection of thioacetamide (500 mg/kg) in rats. Severe liver injury was demonstrated 24 h after the insult and was associated with reduced renal arterial blood flow and glomerular filtration rate and sodium retention. The response to a saline volume expansion challenge was inhibited. As with the other models, 8-phenylpheophylline improved urine production. To specify the adenosine receptor subtype, selective adenosine A_1 and A_2 receptor antagonists were compared. The selective A_1 antagonist, 8-cyclopentyl-1,3-dipropylxanthine, greatly improved the impaired renal function induced by acute liver injury and this beneficial effect was blunted in rats with liver denervation. In contrast, intravenous administration of the antagonist was only effective at higher doses, thereby confirming that the adenosine receptor antagonist was acting on the liver and not directly on the kidney. The adenosine A_2 agonist was without impact on the renal function [270].

Although both the chronic and acute liver disease models clearly demonstrated an adenosine-dependent hepatorenal reflex impairment of renal function, the relationship to intrahepatic portal flow in diseased livers cannot be assumed. Adenosine concentrations in the space of Mall can be elevated by reduced portal flow or intrahepatic vascular shunting, but it is equally possible that adenosine levels could be elevated independent of blood flow, secondary to hepatic inflammation [290,292,324] or by a decrease in the recycling of adenosine through the adenosine kinase pathway [26]. Regardless of the source of increased adenosine in the diseased state, the normal physiology is strongly supportive of a hepatic reflex mechanism by which the liver indirectly affects its own blood flow by adjusting major homeostatic parameters. Those adjustments are pathogenic in the presence of portacaval shunting of blood around the liver. The involvement of this reflex in liver disease suggests a therapeutic approach treating the early renal dysfunction and, perhaps, even the late-stage hepatorenal syndrome through the blockade of intrahepatic adenosine A_1 receptors. Caffeine blocks the hepatic adenosine A_1 but not A_2 receptors (Ming and Lautt, unpublished observation) and can therefore be considered to treat fluid retention associated with reduced hepatic blood flow, including congestive heart failure. A slow release formulation should undergo clinical trial (Chapter 17). The role of the hepatorenal reflex in the homeostasis of hepatic blood flow is discussed in Chapter 16.

• • • •

CHAPTER 14

Integrative Hepatic Response to Hemorrhage

The hepatic response to hemorrhage is an elegant example of interactive homeostatic mechanisms functioning to compensate for a severe homeostatic disturbance. In this chapter I will briefly review the hepatic vascular responses that have been discussed in several chapters in this monograph. In addition, I will incorporate the metabolic responses that regulate glucose metabolism and are intimately connected with both vascular and metabolic emergency responses.

A rapid hemorrhage leads to an immediate decrease in hepatic blood volume. The liver compensates for approximately 20% of either an increase or a decrease in circulating blood volume. Hepatic blood volume is able to be actively decreased (unstressed volume) by high levels of vasopressin or angiotensin (small effects) or by blood-borne catecholamines (small effect) or by activation of hepatic sympathetic nerves. Hepatic sympathetic nerve stimulation represents the most powerful regulator of hepatic blood volume. However, if all of these active regulators are eliminated by removing the pituitary and adrenal glands and kidneys and by hepatic denervation, the ability of the liver to respond to hemorrhage is impaired only slightly, as the passive effect of decreased portal flow and blood pressure result in a reduction in stressed blood volume [204]. Hemorrhage can result in expulsion of up to 50% of the entire blood content of the liver, which is equivalent to a 6–7% infusion of total blood volume. Hemorrhage appears to result in hepatic blood volume changes primarily through changes in stressed volume, as reducing portal inflow to the same extent as is produced during hemorrhage results in a similar level of capacitance response. Furthermore, the change in unstressed volume caused by nerve stimulation is unaltered if stressed volume is already reduced by hemorrhage. Thus, the effects of changes in stressed and unstressed volume can be additive (Figure 14.1).

The major decrease in hepatic blood flow that occurs in response to hemorrhage is a result of arterial vasoconstriction of the splanchnic organs feeding the portal venous flow. This dramatic reduction in portal flow activates the hepatic arterial buffer response, which leads to dilation of the hepatic artery and thereby protects the oxygen supply of the liver [227].

The glycogenolytic response of the liver is also a major component of the homeostatic compensatory responses. Blood loss results in a rapid activation of glycogen breakdown in the liver

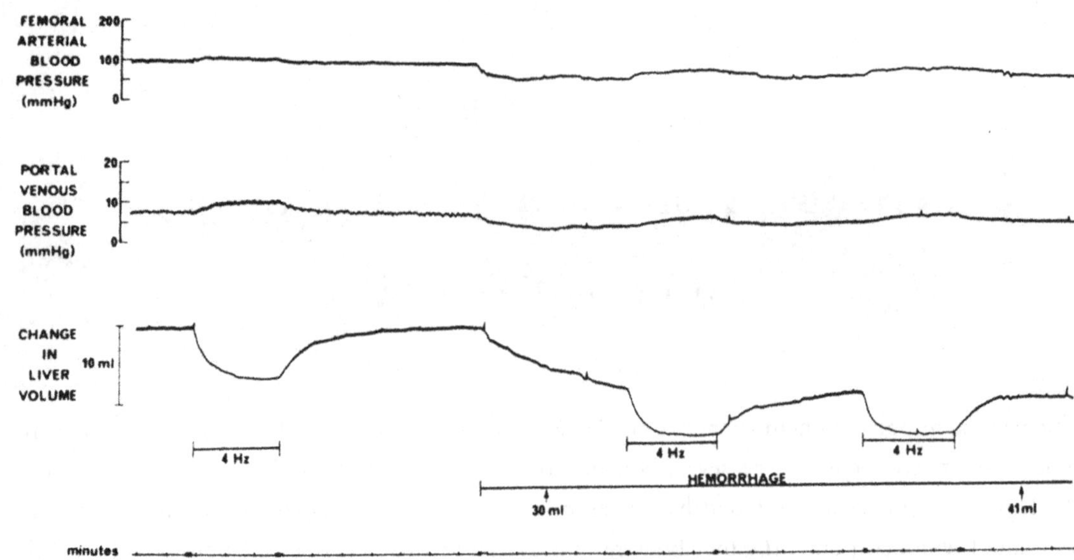

FIGURE 14.1: Effects of hepatic nerve stimulation in one cat before and after hemorrhage. After the reduction in stressed volume after hemorrhage, change in unstressed volume produced by nerve stimulation is almost unchanged. Effects of passive and active mobilization of blood volume are additive. Reprinted from Greenway CV, Lautt WW. Hepatic circulation. In: *Handbook of Physiology, The Gastrointestinal System, Motility and Circulation.* Bethesda: Am Physiol Soc, Section 6, Volume 1, Part 2, Chapter 41, pp. 1519–1564, 1989. (This figure from publication Handbook of Physiology, The Gastrointestinal System, Motility and Circulation is reproduced with permission of publisher Am Physiol Soc).

leading to blood glucose increasing from a normal range of 100 mg% to as high as 800 mg%. This response is mediated by a redundant control system [210]. The system is redundant in that the hepatic glycogenolysis can be equally activated by the hepatic sympathetic nerves or the adrenal secretions of catecholamines. Elimination of either regulator produces minor impairment of the glycogenolytic response, whereas elimination of both regulators essentially eliminates the hyperglycemic response to hemorrhage. This demonstration is extremely important in that it is an example for the need for caution in interpretation of ablation studies. Lack of impact of elimination of one putative regulator does not allow the conclusion to be made that the regulator is not a significant factor. As long as either the hepatic nerves or the adrenal glands are intact, the response is maintained.

In addition to this neural homeostatic role, the hepatic nerves decrease peripheral insulin sensitivity, thereby preserving the elevated glucose levels as a fuel to be transported to the brain and eyes in high concentration in spite of the reduced blood supply. This mechanism appears to be mediated by hepatic nerves that release somatostatin within the liver and cause blockade of insulin-induced release of a hepatic insulin-sensitizing substance (HISS). Insulin secretion is also blocked, also apparently by somatostatin. The discussion of the HISS hypothesis and the hepatic metabolic regulation of peripheral insulin resistance is beyond the purview of this monograph but has been recently reviewed [202].

The increase in hepatic glucose concentration contributes to increased extracellular osmotic fluid pressure, thereby drawing fluid from the large intracellular fluid compartment. Intracellular water accounts for approximately two thirds of total body water and as much as 1 liter has been estimated to be reabsorbed into the plasma compartment over a 1- to 2-h period in humans [154].

Activation of the hepatic sympathetic nerves therefore results in both a dramatic vascular response and metabolic response. If the vasoconstriction at the arterial or portal venous resistance sites increases regional shear stress, nitric oxide is released from the endothelial cells and results in inhibition of the vasoconstriction but potentiation of the hyperglycemic response to catecholamines (see Chapters 5, 9, and 11).

The sympathetic nerves cause vasoconstriction in the hepatic artery and portal vein. If portal blood flow has decreased in response to the hemorrhage, vasoconstriction at the portal site may produce little or no change in portal pressure as the increase in resistance compensates for the decrease in flow. Vasoconstriction in the hepatic artery occurs rapidly and dramatically in response to either norepinephrine or electrical nerve stimulation. The vasoconstriction reaches a peak within 2 min and then may undergo vascular escape so that blood flow returns toward control levels. This vascular escape from neurogenic vasoconstriction is a result of shear stress-induced release of nitric oxide (Chapter 11). If blood pressure is reduced in response to hemorrhage, the shear stress-induced inhibition of vasoconstriction will not occur. However, the second major intrinsic vascular regulator, adenosine, in situations of reduced total hepatic blood flow, can result in suppression of the vasoconstriction in the hepatic artery with little or no inhibitory effect on sympathetic nerve-induced constriction of the portal vein or capacitance vessels (Chapter 10).

Thus, the hepatic autonomic nerves play a major role in the integrated hepatic response to hemorrhage. The sympathetic nerves constrict the hepatic artery, portal vein, and hepatic capacitance vessels, and trigger glycogenolysis. Flow escape in the artery can occur as a result of either shear stress-induced nitric oxide action (less likely because arterial pressure is reduced) or activation of the hepatic arterial buffer response, secondary to reduced portal flow. The hepatic parasympathetic control of skeletal muscle insulin sensitivity is blocked by intrahepatic somatostatin release, thereby contributing to hemorrhage-induced hyperglycemia.

• • • •

CHAPTER 15

Blood Flow Regulation of Hepatocyte Proliferation

In this chapter I discuss the hypothesis, first proposed in 1997 [371], that changes in portal blood flow lead to shear stress-dependent changes in hepatic nitric oxide production, which serves as the initial trigger for the activation of a complex cascade of events leading to cellular proliferation, in the case of elevated portal flow, or apoptosis, in the case of reduced portal flow. The hemodynamic consequence of partial hepatectomy is the trigger for liver regeneration that occurs at an explosive rate.

Before 1954, there were a number of studies that were compatible with the hypothesis that hepatic blood flow regulated liver cell mass, but a few poorly conducted and improperly interpreted studies led to a rapid consensus that hepatic blood flow was not a significant regulator of liver mass. However, in a review on hepatic circulation, Greenway and Lautt [114] suggested that the coincidence of enzyme induction and elevated portal flow, which had been interpreted to suggest that an increase in hepatic metabolism and liver volume led to an increase in portal flow, was inconsistent with several clearly defined studies demonstrating that the liver cannot directly control portal blood flow. We suggested that the data were better accounted for by blood flow controlling liver cell mass rather than liver cell mass controlling blood flow.

The liver is well recognized to have a unique ability to rapidly regenerate. Perhaps even the ancient Greeks knew of this remarkable ability because the legend of Prometheus describes the wrathful punishment by Zeus, for the sin of revealing the secret of fire to mankind, by the unique torture of the chained Prometheus having his liver plucked out by an eagle by day only to have it regenerated by night, thus perpetuating his torment indefinitely. Although the extent of hepatic liver regeneration is exaggerated by this legend, it remains a striking observation that after a two-thirds partial hepatectomy in rats, full restoration of liver volume can be obtained within approximately 1 week and 50% of the recovery occurs within 48 h. In a review of hepatic regeneration, Michalopolous and DeFrances [267] indicated that, despite more than 100 years of research, the trigger of liver regeneration remained unknown and that the discovery of this trigger would be akin to the big bang theory of evolution of the universe.

For a finite event to be proposed as a trigger for the regeneration cascade, the event must occur immediately after the partial hepatectomy and serve as a trigger for the entire cascade. A dramatic hemodynamic stimulus occurs at the time of surgical removal of liver lobes or selective ligation of portal lobar veins. With the classic model of a two-thirds partial hepatectomy, all of the portal blood flow is forced to pass through the remaining liver mass, thereby increasing the flow-to-mass ratio to three times normal levels. Shear stress is defined as the viscous drag at the surface of endothelial cells created by adjacent blood flow [43,159]. The amount of shear stress is proportional to the blood flow through a vessel and the inverse of the cube of the vessel radius [43]. Thus, an increase in blood flow velocity causes an increase in shear stress on the endothelial cells in the remnant liver. Increased flow, generating shear stress, causes release of nitric oxide (NO) from endothelial cells [166]. It is this release of NO, in response to an increase in shear stress in the liver, that we hypothesize triggers the liver regeneration cascade, including production of growth factors and cytokines.

Shear stress in hepatic circulation is reflected as an increase in portal venous pressure. Theoretical calculations and experimental data indicate that changes in portal venous pressure reflect changes in shear stress. Macedo and Lautt [252] showed that prevention of elevation of portal pressure, by reducing portal blood flow during vasoconstriction, prevented nitric oxide release. However, if portal pressure was allowed to rise, by holding the flow steady, nitric oxide caused compensatory vasodilation. If vascular perfusion pressure rises, either as a result of increased flow or increased resistance, the elevated perfusion pressure reflects increased shear stress in venous as well as arterial vascular beds [250,252]. Portal pressure increases immediately after partical hepatectomy or selective portal branch ligation [129,364] and remains elevated until liver mass is restored [364]. Also, nitric oxide is released after partial hepatectomy [138,291], which is compatible with the hypothesis that NO, released in response to an increase in shear stress, triggers the liver regeneration cascade.

The first studies to test this shear stress/nitric oxide hypothesis [371,372] were based on previous observations that a wide range of hepatic proliferating factors appeared in the plasma of animals that had been subjected to a partial hepatectomy. We developed a bioassay to detect the presence of proliferating factors, using the ability of plasma from a rat with partial hepatectomy to stimulate hepatocyte proliferation in vitro. Blood removed from animals that had been subjected to a two-thirds partial hepatectomy showed maximal proliferative stimulation from samples drawn 4 h after the partial hepatectomy. The nitric oxide synthase antagonist, L-NAME, was administered to prevent shear stress-induced stimulation of nitric oxide production. Blood from these animals showed no proliferative activity [371]. The response was restored by provision of a nitric oxide donor to the liver [373].

Subsequent studies evaluated the earliest and latest stages of the regeneration cascade. At the early stage, we used the expression of an immediate early gene that had previously been shown to reach a peak activation 15 min after partial hepatectomy and was dependent on the degree of partial hepatectomy performed [280]. C-fos activation was shown to occur in the remnant liver after partial hepatectomy and not in sham-operated animals [337]. C-fos mRNA expression was prevented by blocking hepatic nitric oxide synthase activation and by blocking prostaglandin production, both of which are regulated by shear stress. Activation of c-fos was inhibited by blockade of nitric oxide synthase or cyclooxygenase and could be reversed in both cases by administration of nitric oxide donors and the prostaglandins, PGE_2 and PGI_2, suggesting that there is an interaction between nitric oxide and prostaglandins in triggering the liver regeneration cascade [338]. The late response was quantified from liver mass restoration determined 48 h after the partial hepatectomy. The phosphodiesterase V antagonist, zaprinast, the nitric oxide donor, SNAP, and PGI_2 potentiated early c-fos mRNA expression and 48-h hepatocyte mass restoration. Cyclic GMP was a likely mediator of the NO effect.

The relationship between hepatic blood flow and regulation of hepatocyte proliferation was strongly supported by the demonstration that prevention of shear stress after partial hepatectomy blocked the activation of the regeneration cascade. Occlusion of the superior mesenteric artery decreases hepatic blood flow by approximately two thirds; a two-thirds partial hepatectomy delivers the entire portal flow to one third of the normal liver mass, thereby increasing hepatic blood flow per remaining liver mass by three times. Therefore, occlusion of the superior mesenteric artery after a two-thirds partial hepatectomy should prevent the development of shear stress in the remnant liver. This was shown by a lack of activation of c-fos in this model [337].

Selective ligation of portal lobar veins leads to decreased portal flow in the ligated lobes with elevated flow to the unligated lobes. Liver volume adjusts so that flow per unit liver weight is restored after 1 week by hypertrophy of the unligated lobes and atrophy of the ligated lobes [318]. The selective ligation of the left branch of the portal vein resulted in increased portal flow to the unligated two thirds of the liver and led to similar elevation in portal pressure as was achieved by two-thirds partial hepatectomy of the same lobes, thus indicating similar elevations of shear stress. The resultant elevations in c-fos in the unligated lobes and the appearance of proliferating factors in plasma were similar to what was seen after surgical removal of the ligated lobes and could be blocked by nitric oxide synthase antagonists [337] (Figure 15.1). This study confirmed that the trigger for regeneration was hemodynamic in nature and regulated by nitric oxide and was not dependent on reduction of liver parenchymal cell mass.

The relationship of portal pressure (shear stress) to the triggering of regeneration was also shown by Sato et al. [333] who suggested that there was an upper limit to a beneficial effect of elevated portal pressure. A 90% partial hepatectomy raised portal pressure to an extent that was suggested to account for a lesser degree of effective regeneration. c-fos activation 15 min after partial hepatectomy also increased in proportion to the degree of ablation, but the 90% hepatectomy was less effective [280]. Maintained hyperdynamic portal circulation, seen after human liver transplant, was also associated with more rapid liver regeneration [74].

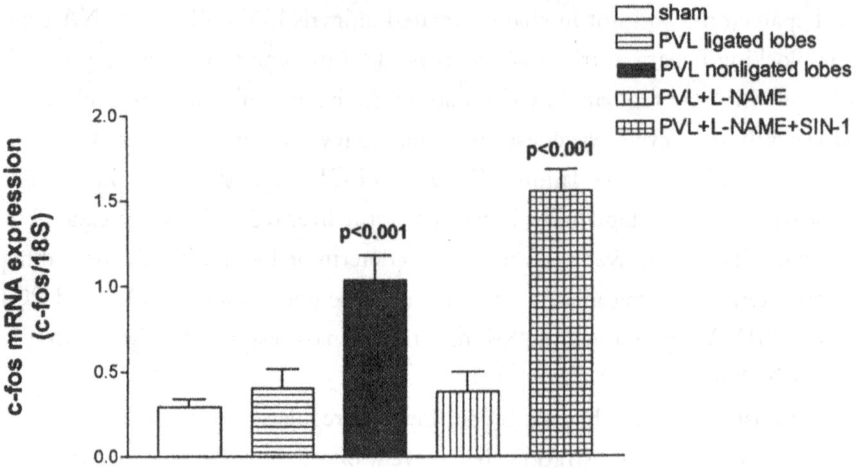

FIGURE 15.1: Hepatic *c-fos* mRNA expression after selective portal vein ligation (PVL). Hepatic *c-fos* mRNA expression increased significantly in the nonligated lobes 15 min after PVL compared to that in sham-operated animals. This increase in *c-fos* mRNA expression was inhibited by L-NAME, and the inhibition was reversed by the NO donor SIN-1. These results suggest that *c-fos* mRNA expression increases in response to hemodynamic changes in hepatic blood flow. Also, *c-fos* mRNA expression did not increase in the ligated lobes after PVL, suggesting that *c-fos* mRNA is selectively expressed in the nonligated lobes in response to a hemodynamic change rather than in response to surgical trauma. *c-fos* expression is a good index for the initiation of the liver regeneration cascade. Reprinted from Nitric Oxide 5(5), Schoen JM, Wayne HH, Minuk GY, Lautt WW. Shear stress-induced nitric oxide release triggers the liver regeneration cascade, pp. 453–464, 2001, with permission from Elsevier. (This figure from publication Nitric Oxide is reproduced with permission from Elsevier).

These hemodynamic relationships to shear stress and liver volume do not appear to have been studied in liver disease, but the presence of portacaval shunts and altered intrahepatic hemodynamics could be a major cause of reduced hepatic regenerative capacity in diseased states.

The relationship between hepatic sinusoidal blood flow, shear stress-induced regulation of nitric oxide, and regulation of hepatocyte parenchymal mass is anticipated to act at a much more subtle level than was demonstrated for the experimental manipulations, primarily focusing on surgical removal of a large portion of the liver or vascular changes of a large magnitude. It is anticipated that small changes in sinusoidal flow and resultant shear stress will result in a dynamic homeostatic interaction between cellular proliferation and apoptosis. This would account for why the superficial physical structure and shape of the liver is so readily able to accommodate to intra-abdominal objects, including gastric tumors.

• • • •

CHAPTER 16

Multiple Mechanisms Maintaining a Constant Hepatic Blood Flow to Liver Mass Ratio

In this chapter, I will discuss how several unique characteristics of the hepatic circulation interact to adjust the ratio of total hepatic blood flow to hepatic parenchymal cell mass. These regulatory processes act over different time scales and with different consequences, but the ultimate homeostatic role is to maintain constancy of blood flow relative to the metabolic machinery of the liver. This is important for homeostatic stability, as clearance of many drugs and hormones is blood flow-dependent and a relatively constant rate of removal of hormones from the plasma is necessary to allow fine control of blood levels by the endocrine glands.

The importance of constant hepatic blood flow to endocrine homeostasis has been discussed in Chapter 5, hepatic capacitance roles in Chapter 4, the effect of neuromodulators in Chapter 11, the role of hepatic afferent control over renal efferent function in Chapter 13, and the effect on liver regeneration in Chapter 15. To minimize repetition, only the conceptual aspects of these regulatory systems will be considered in this chapter.

16.1 OVERVIEW

There are multiple interrelated mechanisms that act acutely and chronically to maintain a constant hepatic blood flow-to-liver mass ratio. Maintenance of hepatic blood flow is made more complex by the unique characteristics of the hepatic vascular bed. The liver receives approximately 25% of the entire cardiac output and three quarters of that blood flow is provided to the liver through the portal venous drainage from the stomach, intestines, spleen, pancreas, and visceral fat. The liver is not capable of directly controlling portal blood flow. Yet, as will be discussed, the liver can have very significant indirect effects to regulate portal blood flow via mechanisms impacting on blood flow to the splanchnic organs that drain into the portal vein.

The first mechanism is a simple physical consequence of the very high vascular capacitance (blood volume) and compliance (change in hepatic blood volume per unit change in intrahepatic pressure). A decrease in portal blood flow leads to a passive decrease in intrahepatic pressure and a passive expulsion of blood from the large hepatic reservoir into the central venous system. This increase in venous return leads to increased cardiac output that, in turn, leads to elevated blood flow in the splanchnic arteries that feed the portal venous bed, thus at least partially correcting the initial flow deficit.

At the same time, the reduced portal flow activates the hepatic arterial buffer response (HABR) secondary to reduced washout of adenosine from the space of Mall, which surrounds the terminal branches of the portal vein and hepatic artery before they drain into the hepatic sinusoids. Adenosine appears to be secreted at a constant rate into the space of Mall, with the local concentration of the potent vasodilator being regulated by the rate of washout into the portal blood. By this mechanism, reduced portal flow leads to accumulation of adenosine and hepatic arterial dilation, thereby serving to buffer the impact that changes in portal flow have on total hepatic blood flow. This is the mechanism of the HABR.

The accumulated adenosine also activates sensory nerves in the liver, which results in activation of a hepatorenal reflex. This reflex leads to reduced renal output and fluid retention, thereby elevating blood volume, venous return, cardiac output, and splanchnic blood flow. The elevated adenosine level that occurs in response to reduced portal flow leads to rapid responses. The hepatic artery is dilated within seconds and the response is well maintained. The hepatorenal reflex is also activated immediately, but the renal fluid retention has cardiovascular consequences only after a much longer time scale, after fluid retention becomes significant.

Hepatocyte proliferation is also rapidly activated by a hemodynamic mechanism related to vascular shear stress but is modulated over a longer period of time. Hepatic vascular shear stress regulates nitric oxide and prostaglandin release to trigger a cascade leading to hepatocyte proliferation or apoptosis. By this mechanism, hepatic cell mass is adjusted to maintain a constant ratio to the mean chronic hepatic blood supply.

Nitric oxide and adenosine play distinct roles to inhibit vasoconstrictors: nitric oxide acts if local hepatic arterial or portal venous constriction raises shear stress; adenosine acts if hepatic blood flow is reduced.

The focus in this review will be on concepts. Detailed references are available in the original articles and reviews. Figure 16.1 represents a conceptual summary.

16.2 HEPATIC COMPLIANCE

The liver accounts for 12% of the total blood volume and half of that blood can be rapidly expelled from the liver in response to both active and passive influences, thus giving the liver a dramatic role as a blood volume reservoir.

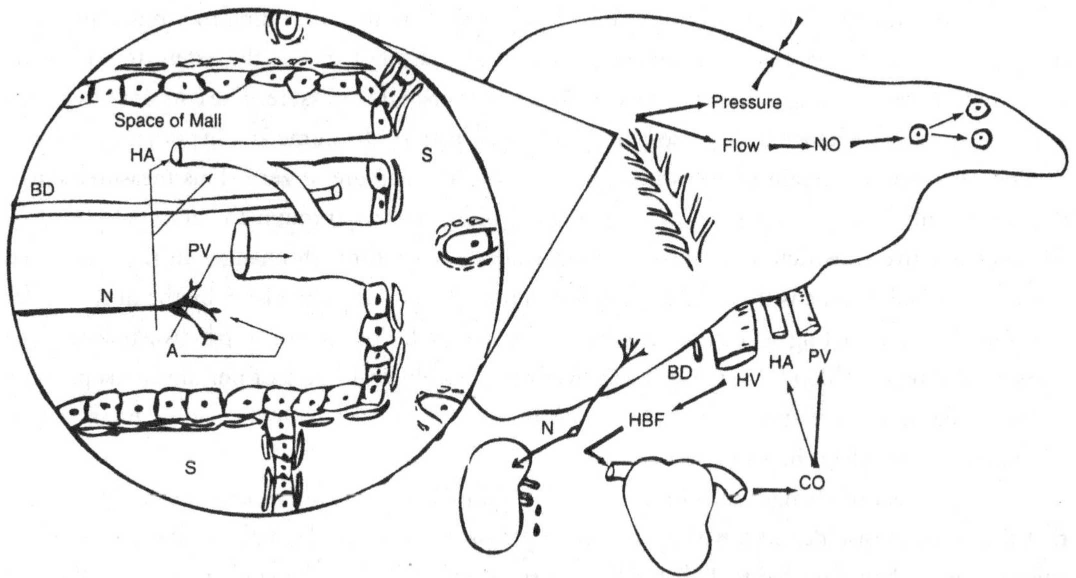

FIGURE 16.1: The space of Mall is a minute fluid space surrounding the hepatic arterioles (HA), bile ductules (BD), portal venules (PV), and sensory nerves (N) receptive to adenosine. Adenosine (A) is secreted into the space of Mall, and its concentration is regulated by the rate of washout from the space of Mall into the blood vessels. A reduction in blood flow leads to immediate increase in adenosine levels, which acts on the HA causing vasodilation and inhibition of vasoconstrictors. Adenosine does not modulate the PV. Portal and arterial blood flow wash adenosine from the space of Mall, thus controlling HA vascular tone and blood flow. Adenosine also acts on hepatic sensory nerves that serve to reflexly regulate renal fluid retention, and therefore, indirectly, cardiac output (CO) and hepatic blood flow (HBF). Outside the space of Mall, a highly vascular and distensible liver responds to changes in intrahepatic distending blood pressure to regulate the large hepatic blood reservoir. The hepatic cells (probably endothelial) respond to blood flow that is sensed via shear stress. Vascular shear stress results in the release of nitric oxide (NO). NO dilates the resistance vessels in the HA and PV and blocks vasoconstrictors. NO causes vascular escape from vasoconstriction in the HA. NO also activates the machinery of hepatic proliferation serving as the trigger to adjust the functional cell mass of the liver. HV, hepatic vein; S, sinusoid. Reproduced with permission from Lautt WW. Regulatory processes interacting to maintain hepatic blood flow constancy: vascular compliance, hepatic arterial buffer response, hepatorenal reflex, liver regeneration, escape from vasoconstriction. *Hepatol Res* 37(11): pp. 891–903, 2007, published by Wiley-Blackwell, UK. (This figure from publication Hepatol Res is reproduced with permission of publisher Wiley-Blackwell, UK).

The determinants of the volume of blood in an organ are the distending transmural pressure, the compliance of the vessels, and the unstressed volume. Compliance is the extent to which the volume of the vessel changes in response to a change in transmural pressure. The concept of stressed and unstressed capacitance is discussed in Chapter 4. Unstressed volume is a hypothetical volume of blood that would remain within the organ at a vascular pressure of zero. This measurement is obtainable only through extrapolation of pressure–volume curves through the zero pressure axis. All known active constrictors of hepatic blood volume do so through changes in the unstressed volume. Stressed volume is the volume of blood in the organ due to distension by the intravascular transmural pressure acting against the compliant vascular bed. The relationship between distending pressure and hepatic blood volume is linear over the physiological range of portal venous pressure [121]. Portal venous pressure provides a good estimate of the intrahepatic distending pressure that acts upon the compliant hepatic vascular bed.

The impact of changes in portal flow on portal pressure is made more complex by the fact that the venous resistance sites within the liver are passively distensible, with the intrahepatic vascular resistance being related to 1/distending pressure cubed. Thus, a large change in portal blood flow can be partially compensated for by what we have referred to as portal pressure autoregulation [219]. The distensible resistance sites passively change to minimize changes in portal pressure so that doubling portal flow, or reducing it by 50%, results in changes of only a few mmHg in portal pressure. Nevertheless, the extremely compliant vascular bed of the liver responds rapidly and dramatically to small changes in distending pressure, with an increase in distending pressure of 8 mmHg resulting in doubling of the hepatic blood volume [212]. Partial occlusion of the superior mesenteric artery in dogs led to a reduction of portal venous pressure of 2.8 mmHg and a rise in cardiac output of 19% [125].

A decrease in portal blood flow results in a decrease in portal pressure and a passive expulsion of blood from the liver into the central venous compartment. Depending on the mechanism that caused the reduction in portal flow, the hepatic blood volume response serves either to maintain the venous return (preload to the heart) or to actually increase it, thus maintaining or elevating the cardiac output, which, in turn, results in increased flow to the splanchnic arteries and a partial correction of the portal flow.

The role of hepatic compliance in diseased livers is not clear. The hepatic blood volume response to hemorrhage was dramatically decreased in a 14-day chronic bile duct ligation model of liver disease in cats. However, hepatic compliance was, unexpectedly, not affected despite the presence of severe biliary hyperplasia, portal tract distortion, and fibrosis. The reduction in hepatic vascular response to hemorrhage appeared to be accounted for by dysfunction of hepatic sympathetic nerves as the direct response to nerve stimulation was severely impaired, whereas the response to infused norepinephrine was well maintained [335] (stressed volume was not altered, unstressed volume was impaired).

Although research has not been specifically carried out to examine the stressed and unstressed volume responses in the more severely diseased liver, it seems reasonable to anticipate that a cirrhotic liver would be less distensible and therefore play a reduced role in this disease state. Similarly, when cirrhosis results in extensive portacaval shunting, any change in portal inflow is reflected as changes in the outflow via the portacaval shunt not as change in intrahepatic blood flow. The liver would not be exposed to the changed flow or pressure and portal inflow would not impact on hepatic capacitance, thus eliminating this mechanism as a regulator of hepatic blood flow.

16.3 THE HABR

The liver is not master of its own blood supply. The liver must accept and accommodate two thirds of its blood flow arriving from the portal venous effluent of the splanchnic organs. The portal flow arising in the highlands of the splanchnic organs is individually regulated by these organs, so that the total blood flow entering the splanchnic arteries at a blood pressure of 120 mmHg is decreased to ≈3–10 mmHg in the portal vein before entry to the liver. The liver cannot control portal blood flow (Chapters 5 and 11). A reduction in portal blood flow leads to a rapid dilation of the hepatic artery explained by the mechanism of the HABR (Chapter 5).

Briefly, the HABR is accounted for by the following mechanism. The hepatic artery and the portal vein undergo progressive parallel divisions that eventually travel as their terminal branches through the small space of Mall, which is surrounded by a limiting plate of hepatocytes. Adenosine appears to be produced at a constant rate, independent of oxygen supply or demand, and is secreted into the space of Mall where it serves as a powerful dilator of the hepatic artery. The concentration of adenosine is regulated by the rate of washout into the blood vessels that pass through the space of Mall. According to this theory, a decrease in portal flow results in a reduced washout of adenosine and the accumulated adenosine concentration results in dilation of the hepatic artery, thus partially compensating for the decrease in portal blood flow.

16.4 THE HEPATORENAL REFLEX

The hepatorenal reflex is described in Chapter 13. Briefly, the hypothesis is that reduced hepatic blood flow leads to compensatory increase in renal fluid retention, with a subsequent increase in circulating plasma volume, increased venous return, and increased cardiac output. The increase in cardiac output results in increased portal blood flow thereby, at least partially, correcting the decrease in intrahepatic blood flow. Aspects of this hypothesis are discussed in detail in Chapter 13 but include the following: adenosine administered to the liver through portal venous infusion results in decreased renal excretion of salt and water; adenosine receptor antagonists of the A_1 subtype block the adenosine mediated renal retention; decrease in hepatic portal blood flow in the absence of systemic changes in central venous pressure or arterial pressure result in renal fluid retention; denervation of the liver or kidney result in elimination of the fluid retention produced by adenosine or reduced portal blood flow; chronic liver disease induced by thioacetamide administration results in reduced renal salt and water excretion and reduced ability to rapidly excrete a saline load; adenosine A_1 receptor antagonists, including caffeine, improve renal function in a chronic liver disease model. In total, these series of progressive studies are compatible with the hypothesis that a decrease in portal flow results in an increase in intrahepatic adenosine concentration that activates hepatic afferent nerves, responding through A_1 adenosine receptors, leading to a reflex inhibition of renal salt and water reabsorption. This response is a normal physiological response to regulate total hepatic blood flow at a constant level, but is apparently also the mechanism of severe fluid retention in chronic liver disease, and is thus proposed to account for the homeostatic fluid imbalance seen with chronic liver disease.

16.5 MODULATION OF VASOCONSTRICTORS BY ADENOSINE AND NITRIC OXIDE

Vasoconstriction of the hepatic artery, whether produced by sympathetic nerve activation or delivery of vasoconstrictors to the resistance vessels through the circulation, leads to reductions in hepatic arterial flow. The portal circulation, however, responds to constrictors by an elevation in portal pressure with portal flow remaining constant, as portal flow is regulated by the outflow of the extrahepatic splanchnic organs. Nitric oxide and adenosine play an additional protective role tending to minimize the impact of vasoconstrictors on hepatic blood flow.

The role of nitric oxide in modulating vasoconstrictors in the liver appears to be dependent on shear stress [226]. If vasoconstriction leads to increased shear stress (flow held steady, pressure increasing), nitric oxide release suppresses the constriction, thus protecting the hepatic endothelial cells from shear stress-induced disruption. Blockade of nitric oxide synthase potentiates the response of both the hepatic artery and the portal vein to sympathetic nerve stimulation and norepinephrine infusion. However, if shear stress is not increased during the constriction (flow allowed to decrease, pressure held steady), nitric oxide is not released and blockade of nitric oxide synthase is without impact on vasoconstriction [271]. The hepatic artery responds to vasoconstriction by an initial reduction in blood flow but, in the face of continued stimulation, the flow returns toward baseline by a process referred to as vascular escape. Vascular escape in the hepatic artery is mediated by nitric oxide [271] probably secondary to shear stress induced in the portal vein, as vascular constriction in the portal venules does not reduce portal flow but does increase portal pressure and therefore shear stress.

Adenosine has been shown to antagonize the vasoconstriction of the hepatic artery induced by a range of endogenous constrictors including sympathetic nerves, norepinephrine, angiotensin, and vasopressin [217]. In contrast, adenosine has no significant effect on either the basal portal vascular tone or on the action of vasoconstrictors on the portal vein or capacitance vessels in cats, at doses that significantly modulated the arterial responses [231]. Adenosine may, therefore, play a modest role in suppressing vasoconstriction in the hepatic artery.

Thus, the combination of adenosine and nitric oxide provides the liver with a unique ability to escape from severe vasoconstrictor influences. The role of adenosine is likely only activated if total hepatic blood flow is reduced and results in local accumulation of adenosine at the vascular resistance site. In contrast, local constriction restricted to the liver is antagonized by shear stress/nitric oxide-induced vasodilation. By this mechanism, shear stress is minimized in the portal circulation and the rise in portal pressure is attenuated and the reduction in blood flow in the hepatic artery is antagonized. By these mechanisms, the liver can protect itself from excess vasoconstriction whether blood flow is reduced or shear stress is increased, and whether generalized or regional hepatic vasoconstriction occurs. This is important in the liver in contrast to other vascular beds where vasoconstriction results in regional hypoxia and release of vasodilator substances that inhibit the constriction. This mechanism is not tenable in the liver because vasodilator substances released from hypoxic hepatocytes enter the sinusoidal blood and are swept downstream to the hepatic veins without the opportunity for contact with the resistance vessels upstream.

16.6 BLOOD FLOW REGULATION OF HEPATOCYTE PROLIFERATION

If the combination of the capacitance response, the buffer response, and the hepatorenal reflex do not correct an imbalance in the blood flow-to-liver mass ratio, the liver responds by adjusting the hepatocyte total mass through a hemodynamic mechanism. We first proposed in 1997 [371] that changes in portal blood flow lead to shear stress-dependent changes in hepatic nitric oxide production, which serves as the initial trigger for the activation of a complex cascade of events leading to cellular proliferation in the case of elevated portal flow, or apoptosis in the case of reduced portal flow (see Chapter 15).

Support for the suggestion that the trigger for liver regeneration is related to shear stress-induced release of nitric oxide (and prostaglandins) is demonstrated using a classic model of a two-thirds partial hepatectomy. Blood taken from animals exposed to two-thirds partial hepatectomy shows increased levels of proliferating factors in blood that can be detected using a hepatocyte tissue culture assay. Blockade of NO production blocked the appearance of proliferative activity in the blood [371], and this response was restored by provision of a NO donor to the liver [373]. Expression of the immediate early gene, c-fos, was shown to increase, reaching a peak activation 15 min after partial hepatectomy, and was dependent on the degree of partial hepatectomy performed [280]. C-fos activation occurred in the remnant liver after partial hepatectomy and not in sham-operated animals [337]. C-fos mRNA expression was prevented by blocking hepatic nitric oxide synthase activation and by blocking prostaglandin production, both of which are regulated by shear stress [338].

Liver mass restoration determined 48 h after partial hepatectomy was stimulated by a phosphodiesterase V antagonist (zaprinast). Nitric oxide donors and prostaglandin I_2, which potentiated early c-fos mRNA expression, potentiated the hepatocyte mass restoration determined after 48 h.

In the partial hepatectomy model, the shear stress is induced by virtue of the total portal blood flow being forced to pass through the reduced vascular bed. However, if the blood flow in the superior mesenteric artery is decreased so that the reduction in hepatic cell mass is matched by the decrease in portal inflow, c-fos activation is blocked [337]. Selective ligation of portal lobar veins leads to decreased portal flow in the ligated lobes with elevated flow to the unligated lobes. Liver volume adjusts so that flow per unit liver weight is restored after 1 week by hypertrophy of the unligated lobes and atrophy of the ligated lobes [318].

The homeostatic importance of maintaining a relatively constant hepatic blood flow is suggested by the multiple simultaneous regulatory mechanisms of vastly different types. The passive character of the capacitance vessels contrasts with the active hepatic arterial buffer response. The role of adenosine is threefold with adenosine accounting for the HABR as well as activating the sensory limb of an autonomic nerve reflex acting on the kidneys. Adenosine further antagonizes vasoconstrictors. This elegant symphony of homeostasis is capped off by the liver cell mass being adjusted by the hepatic circulation through shear stress regulated nitric oxide, which also suppresses vasoconstriction.

• • • •

CHAPTER 17

Pathopharmacology and Repurposing Drugs as a Research Strategy

There is always a tension between the desire to do pure basic curiosity, or discovery-oriented, research and applied research with a definitive clinical output target. Unfortunately, those controlling the research budgets continuously forget history, which demonstrates quite clearly that the major conceptual breakthroughs that lead to applications consistently arise from curiosity based research. The process that I am going to describe should be especially useful for those whose interest is to develop therapeutics to rapidly enter the market. However, it is intended more as an example of one method of organizing an approach to science. It is a practical reality of the funding process that the referees who are making judgments about the utility of funding a particular research project must generally be convinced that there is some hope of practical application to result. The politicians insist upon it. The real issue that should have to be sold to these people is whether the knowledge derived will be novel and significant, without an immediate and obvious extrapolation to how this new knowledge will rock the foundations of the scientific and business world. Nevertheless, the practical reality dictates that we select research topics that have a high likelihood of being funded and that have a high likelihood of providing some great value either through basic knowledge or through application to human health to be fundable.

I will provide one illustrative example that relates specifically to the hepatic vasculature.

To use this approach, it is assumed that there will be a therapeutic target identified at the outset, usually as a result of prior experimentation or based on the literature. The research direction can then be categorized sequentially from physiology, to pharmacology, to pathophysiology, to pathopharmacology, to drug repurposing, to clinical trials. This approach is often not amenable to reductionist (genetic, molecular, cellular) science but is, rather, dependent on whole animal integrative physiology, pharmacology, and pathology.

Although these concepts require more extensive discussion than is appropriate for this chapter, the example will provide a working model.

17.1 THE PATHOLOGY, THE HEPATORENAL SYNDROME

Patients that die of hepatic liver disease die in renal failure (see Chapter 13). Chronic liver disease is associated with massive accumulation of circulating blood volume and formation of ascitic fluid that can reach volumes of 20 liters in the abdomen of a victim of chronic cirrhosis. The debate as to the mechanism and the therapeutic approach to treat the hepatorenal syndrome has resulted in numerous symposia and contrasting controversial approaches. When searching for a new paradigm to explain a significant pathology or observation, it is important to identify the anomalies in the current accepted or alternate paradigms. It was generally accepted that the renal dysfunction in liver disease was not secondary to a renal dysfunction, per se, but that it was some regulator of renal function that was dysfunctional. Denervation of the liver or kidneys resulted in reduction of the renal fluid retention. The general consensus was that hepatic baroreceptors were responding to hepatic or portal venous hypertension and stimulating reflex sympathetic nerves to the kidney, resulting in renal fluid retention. However, the logic of that explanation implied a physiological positive feedback system whereby an increase in portal blood flow, for instance, which would result in an increase in intrahepatic pressure, would result in activation of sympathetic nerves to the kidneys and fluid retention, which would result in increased circulating blood volume and increased cardiac output and increased portal blood flow, with further elevation in the pressure that was proposed as the afferent limb in this hepatorenal reflex. A positive feedback physiological homeostatic mechanism appeared highly unlikely; therefore, we suggested the unlikely possibility that what was being sensed by the hepatic nerves was blood flow rather than blood pressure. By this model, any disease process that resulted in a reduced portal blood flow into the liver would result in stimulation of the afferent nerves and activation of the hepatorenal reflex. For this hypothesis to be considered viable, it was necessary to propose a specific physiological mechanism by which changes in blood flow could be detected by afferent nerves. Based on our studies related to the hepatic arterial buffer response, we knew that a reduction in portal blood flow would result in an elevation of adenosine in the space of Mall in exactly the region where hepatic sensory nerves have been shown to arise. Because adenosine had been shown to activate sensory nerves in the heart and carotid artery, it seemed feasible that adenosine could also activate sensory nerves in the liver. To test this hypothesis, we administered adenosine into the portal vein in rats and measured renal fluid and electrolyte output. Adenosine administration was intended to mimic the effect of reduced portal blood flow and did result in a decrease in renal fluid and electrolyte output that could be blocked by denervation of the liver, or of the kidney, or by administration of a nonselective adenosine receptor antagonist to the liver. The physiological regulation therefore seemed to be feasible. Additional studies showed that reductions in portal blood flow also produced renal retention that could be blocked by hepatic denervation or adenosine receptor blockade.

Thus, a normal physiological process was reasonably well defined and pharmacological tools were developed.

Pathophysiology and pathopharmacology often overlap. A chronic liver disease model was created using the hepatotoxin thioacetamide and demonstrated to show the anticipated decrease in renal function. However, to have an acute index of renal function, we used an intravenous saline volume stimulus to activate renal fluid and solute excretion. This method proved an extremely valuable tool. The renal dysfunction produced by the chronic liver disease model was similar to that seen with either intraportal adenosine administration or by shunting portal blood around the liver. The pharmacology predicted that the disease state involved adenosine stimulation of hepatic sensory nerves and therefore could be corrected by administration of an adenosine receptor antagonist. The logic was that the chronically damaged liver had elevated intrahepatic adenosine levels due to either or both mechanisms. Chronic liver disease results in increased intrahepatic portal venous resistance and portal hypertension with resultant portacaval shunt formation. According to the hepatic arterial buffer response mechanism, shunting of portal blood to the inferior vena cava results in reduced portal inflow to the liver, which results in increased adenosine concentration in the space of Mall. The elevation in adenosine levels in the space of Mall activates afferent nerves arising in that space, thus activating the hepatorenal reflex. However, another potential source of adenosine could arise from the inflammatory or hypoxic response to the liver, whereby ATP breaks down progressively to ADP, AMP, and finally to adenosine. Inflammatory conditions are often associated with increased production of cyclic AMP, which breaks down to adenosine. Application of the adenosine receptor antagonist corrected the baseline fluid retention and restored the ability to respond to an acute saline overload.

The pathopharmacological phase of the investigation included using selective adenosine receptor antagonists to determine which adenosine receptor subtype was involved. From our previous work with adenosine, we knew that the A_2 receptor was primarily involved with vasodilation and was therefore unlikely to be the relevant receptor. In contrast, the A_1 receptor is known to be the mechanism of stimulation of the central nervous system. An A_1 receptor antagonist restored salt and water excretion and the ability to respond to a saline load, whereas A_2 receptor antagonism was without effect.

Drug repurposing then became a consideration. The question became what was the most suitable adenosine A_1 receptor antagonist that was already on the market and had a solid track record of effects and toxicities. Because of our previous studies, we knew that caffeine had virtually no A_2 receptor antagonist effects in the circulation of either the liver or intestine (Chapter 10), and the literature clearly indicated that caffeine had A_1 antagonistic activity. We therefore tested caffeine as a potential as a repurposed pharmaceutical and demonstrated a very clear therapeutic potential.

The disadvantage with the caffeine approach was that the kinetics of absorption and elimination were too rapid to be applied as a useful diuretic. However, modifying the kinetics of caffeine by using a slow-release capsule formulation offers a viable, testable, and patentable drug repurposing that would allow for direct entrance into phase 2 clinical trials.

The pathopharmacology approach to directing a research program is demonstrated by this one example. A similar approach has led to successful drug repurposing for diabetes therapy that is an entirely different story [233].

• • • •

References

[1] Abraham WT, Lauwaars ME, Kim JK, Pena RL, Schrier RW. Reversal of atrial natriuretic peptide resistance by increasing distal tubular sodium delivery in patients with decompensated cirrhosis. *Hepatology* 22: pp. 737–743, 1995.

[2] Agnisola C, Houlihan DF. Some aspects of cardiac dynamics in octopus volgaris (LAM). In: *Physiology of Cephalopod Mollusks Lifestyle and Performance Adaptations*, edited by Portner HO, O'Dor RK, MacMillian DL. Gordon & Breach Publishers, pp. 87–100, 1994.

[3] Alexander WF. The innervations of the biliary system. *J Comp Neurol* 73: pp. 357–370, 1940.

[4] Allman FD, Rogers EL, Caniano DA, Jacobowitz DM, Rogers MC. Selective chemical hepatic sympathectomy in the dog. *Crit Care Med* 10: pp. 100–103, 1982.

[5] Alvaro D, Gigliozzi A, Piat C, Carli L, Bini A, La Rosa T, Furfaro S, Capocaccia L. Effect of *S*-adenosyl-L-methionine on ethanol cholestasis and hepatoxicity in isolated perfused rat liver. *Dig Dis Sci* 40: pp. 1592–1600, 1995.

[6] Amenta F, Cavallotti C, Ferrante F, Tonelli F. Cholinergic nerves in the human liver. *Histochem J* 13: pp. 419–424, 1981.

[7] Andreen M. Inhalation versus intravenous anaesthesia: effects on the hepatic and splanchnic circulation. *Acta Anaesth Scand* 75: pp. 25–31, 1982.

[8] Andreen M, Irestedt L, Zetterstroem B. The different responses of the hepatic arterial bed to hypovolaemia and to halothane anaesthesia. *Acta Anaesth Scand* 21: pp. 457–469, 1977.

[9] Andreoni KA, O'Donnell CP, Burdick JF, Robotham JL. Hepatic and renal blood flow responses to a clinical dose of intravenous cyclosporine in the pig. *Immunopharmacology* 28: pp. 87–94, 1994.

[10] Angus PW, Morgan DJ, Smallwood RA. Review article: hypoxia and hepatic drug metabolism—clinical implications. *Aliment Pharmacol Ther* 4: pp. 213–225, 1990.

[11] Aoki T, Imamura H, Kaneko J, Sakamoto Y, Matsuyama Y, Kokudo N, Sugawara Y, Makuuchi M. Intraoperative direct measurement of hepatic arterial buffer response in patients with or without cirrhosis. *Liver Transpl* 11: pp. 684–691, 2005.

[12] Barrowman JA, Granger DN. Hepatic lymph. In: *Hepatic Circulation in Health and Disease*, edited by Lautt WW. New York: Raven Press, pp. 137–152, 1981.

[13] Bautista AP, D'Souza NB, Lang CH, Spitzer JJ. Modulation of f-met-leu-phe-induced chemotactic activity and superoxide production by neutrophils during chronic ethanol intoxication. *Alcohol Clin Exp Res* 16: pp. 788–794, 1992.

[14] Becker G, Beuers U, Jungermann, K. Modulation by oxygen of the actions of noradrenaline, sympathetic nerve stimulation and prostaglandin $F_{2\alpha}$ on carbohydrate metabolism and hemodynamics in perfused rat liver. *Biol Chem Hoppe-Seyler* 371: pp. 983–990, 1990.

[15] Beckh H, Balks HJ, Jungermann K. Activation of glycogenolysis and noradrenaline overflow in the perfused rat liver during repetitive perivascular nerve stimulation. *FEBS Lett* 149: pp. 261–265, 1982.

[16] Beckh K, Fuchs E, Balle C, Jungermann K. Activation of glycogenolysis by stimulation of the hepatic nerves in perfused livers of guinea-pig and tree shrew as compared to rat: differences in the mode of action. *Biol Chem Hoppe-Seyler* 371: pp. 153–158, 1990.

[17] Beckh K, Otto R, Ji S, Jungermann K. Control of oxygen uptake, microcirculation and glucose release by circulating noradrenaline in perfused rat liver. *Biol Chem Hoppe-Seyler* 366: pp. 671–678, 1985.

[18] Beckman JS, Beckman TW, Chen J, Marshall PA, Freeman BA. Apparent hydoxyl radical production by peroxynitrite: implications for endothelial injury from nitric oxide and superoxide. *Proc Natl Acad Sci USA* 87: pp. 1620–1624, 1990.

[19] Bennett TD, MacAnespie CL, Rothe CF. Active hepatic capacitance responses to neural and humoral stimuli in dogs. *Am J Physiol Heart Circ Physiol* 242: pp. H1000–H1009, 1982.

[20] Bennett TD, Rothe CF. Hepatic capacitance responses to changes in flow and hepatic venous pressure in dogs. *Am J Physiol* 240: pp. H18–H28, 1981.

[21] Betschart JM, Viryi MA, Perera MIR, Shinozuka H. Alterations in hepatocyte insulin receptors in rats fed a choline-deficient diet. *Canc Res* 46: pp. 4425–4430, 1986.

[22] Billiar TR, Curran RD. Kupffer cell and hepatocyte interactions: a brief overview. *J Parent Enteral Nutr* 14: pp. 175S–180S, 1990.

[23] Billiar TR, Curran RD, Stuehr DJ, West MA, Bentz BG, Simmons RL. An L-arginine-dependent mechanism mediates Kupffer cell inhibition of hepatocyte protein synthesis in vitro. *J Exp Med* 169: pp. 1467–1472, 1989.

[24] Blouin A, Bolender RP, Weibel ER. Distribution of organelles and membranes between hepatocytes and non-hepatocytes in the rat liver parenchyma. *J Cell Biol* 72: pp. 441–455, 1977.

[25] Bolognesi M, Sacerdoti D, Bombonato G, Merkel C, Sartori G, Merenda, R, Nava V, Angeli P, Feltracco P, Gatta A. Change in portal flow after liver transplantation: effect on hepatic arterial resistance indices and role of spleen size. *Hepatology* 35: pp. 601–608, 2002.

[26] Bontemps F, Vincent MF, Van-den-Berghe G. Mechanisms of elevation of adenosine levels in anoxic hepatocytes. *Biochem J* 290: pp. 671–677, 1993.

[27] Bosch B, Garcia-Pagan CG. Complication of cirrhosis. I. Portal hypertension. *J Hepatology* 32(Suppl 1): pp. 141–156, 2000.

[28] Boveris A, Fraga CG, Varsavsky AI, Koch OR. Increased chemiluminescence and superoxide production in the liver of chronically ethanol-treated rats. *Arch Biochem Biophys* 227: pp. 534–541, 1983.

[29] Brater DC. Pharmacokinetics and pharmacodynamics in cirrhosis. In: *The Kidney in Liver Disease*, Fourth Edition, edited by Epstein M. Philadelphia: Hanley & Belfus, Chapter 22, pp. 459–477, 1996.

[30] Brauer RW. Liver circulation and function. *Physiol Rev* 43: pp. 115–213, 1963.

[31] Brauer RW, Holloway RJ, Leong GF. Changes in liver function and structure due to experimental passive congestion under controlled hepatic vein pressures. *Am J Physiol* 197: pp. 681–692, 1959.

[32] Bray GP, Tredger JM, Williams R. *S*-Adenosylmethionine protects against acetaminophen hepatotoxicity in two mouse models. *Hepatology* 15: pp. 297–301, 1992.

[33] Bredfeldt JE, Riley EM, Groszmann RJ. Compensatory mechanism in response to an elevated hepatic oxygen consumption in chronic ethanol-fed rats. *Am J Physiol* 248: pp. G507–G511, 1985.

[34] Bremer C, Bradford BU, Hunt KJ, Knecht KT, Connor HD, Mason RP, Thurman RG. Role of Kupffer cells in the pathogenesis of hepatic reperfusion injury. *Am J Physiol* 267: pp. G630–G636, 1994.

[35] Brizzolara AL, Burnstock G. Evidence for noradrenergic–purinergic cotransmission in the hepatic artery of the rabbit. *Br J Pharmacol* 99: pp. 835–839, 1990.

[36] Brown CM, Collis MG. Adenosine A_1 receptor mediated inhibition of nerve stimulation-induced contractions of the rabbit portal vein. *Eur J Pharmacol* 93: pp. 277–282, 1983.

[37] Burchell AR, Moreno AH, Panke WF, Nealon TF Jr. Hepatic artery flow improvement after portacaval shunt: a single hemodynamic clinical correlate. *Ann Surg* 184: pp. 289–302, 1976.

[38] Burnstock G, Crowe R, Wong HK. Comparative pharmacological and histochemical evidence for purinergic inhibitory innervation of the portal vein of the rabbit, but not guinea-pig. *Br J Pharmacol* 65: pp. 377–388, 1979.

[39] Burnstock G, Crowe R, Kennedy C, Torok J. Indirect evidence that purinergic modulation of perivascular adrenergic neurotransmission in the portal vein is a physiological process. *Br J Pharmacol* 81: p. 533, 1984.

[40] Burt AD, Tiniakos D, MacSween RNM, Griffiths MR, Wisse E, Polak JM. Localization of adrenergic and neuropeptide tyrosine-containing nerves in the mammalian liver. *Hepatology* 9: pp. 839–845, 1989.

[41] Burton GW, Ingold KU. Beta-carotene: an unusual type of lipid antioxidant. *Science* 224: pp. 569–573, 1984.

[42] Burton-Opitz R. The vascularity of the liver. I. The flow of the blood in the hepatic artery. *Quart J Exp Physiol* 3: pp. 297–313, 1910.

[43] Busse R, Fleming I. Pulsatile stretch and shear stress: physical stimuli determining the production of endothelium-derived relaxing factors. *J Vasc Res* 35: pp. 73–84, 1998.

[44] Cabrero C, Duce AM, Oritz P, Alemany S, Mato JM. Specific loss of the high-molecular-weight form of *S*-adenosyl-L-methionine synthetase in human liver cirrhosis. *Hepatology* 8: pp. 1530–1534, 1988.

[45] Cantoni GL. Biological methylation: selected aspects. *Annu Rev Biochem* 44: pp. 435–451, 1975.

[46] Cantoni L, Budillon G, Cuomo R, Rodino S, Le Grazie C, Di Padova C, Rizzardini M. Protective effect of *S*-adenosyl-L-methionine in hepatic uroporphyria: evaluation in an experimental model. *Scand J Gastroenterol* 25: pp. 1034–1040, 1990.

[47] Carlei F, Lygidakis NJ, Speranza V, Brummelkamp WH, McGurrin JF, Peitroletti R, Lezoche E, Bostwick DG. Neuroendocrine innervations of the hepatic vessels in the rat and in man. *J Surg Res* 45: pp. 417–426, 1988.

[48] Carmichael FJ, Saldivia V, Isreal Y, McKaigney JP, Orrego H. Ethanol-induced increase in portal hepatic blood flow: interference by anesthetic agents. *Hepatology* 7: pp. 89–94, 1987.

[49] Carneiro JJ, Donald DE. Change in liver blood flow and blood content in dogs during direct and reflex alteration of hepatic sympathetic nerve activity. *Circ Res* 40: pp. 150–157, 1977.

[50] Chagoya de Sanchez V, Hernandez-Munoz R, Yanez L, Vidrio S, Diaz-Munoz M. Possible mechanism of adenosine protection to carbon tetrachloride acute hepatotoxicity. Role of adenosine by-products and glutathione peroxidase. *J Biochem Toxicol* 10: pp. 41–50, 1995.

[51] Chawla RK, Bonkovsky HL, Galambos JT. Biochemistry and pharmacology of *S*-adenosyl-L-methionine and rationale for its use in liver disease. *Drugs* 40: pp. 98–110, 1990.

[52] Child CG. *The Hepatic Circulation and Portal Hypertension*. Philadelphia: W.B. Saunders, 1954.

[53] Clemens MG, Bauer M, Gingalewski C, Miescher E, Zhang J. Hepatic intercellular communication in shock and inflammation. *Shock* 2: pp. 1–9, 1994.

[54] Clemens MG, Bauer M, Pannen BHJ, Bauer I, Zhang JX. Remodeling of hepatic microvascular responsiveness after ischemia/reperfusion. *Shock* 8: pp. 80–85, 1997.

[55] Cohn JN, Khatry IM, Groszmann RJ. Hepatic blood flow in alcoholic liver disease measured by an indicator dilution technique. *Am J Med* 53: pp. 704–714, 1972.

[56] Cohn R, Kountz S. Factors influencing control of arterial circulation in the liver of the dog. *Am J Physiol* 205: pp. 1260–1264, 1963.

[57] Comparini L. Lymph vessels in the liver in man. *Angiologica Basel* 6: pp. 262–274, 1969.

[58] Cousineau D, Goresky CA, Rose CP. Blood flow and norepinephrine effects on liver vascular and extravascular volumes. *Am J Physiol Heart Circ Physiol* 244: pp. H495–H504, 1983.

[59] Cousineau D, Goresky CA, Rose CP, Lee S. Reflex sympathetic effects on liver vascular space and liver perfusion in dogs. *Am J Physiol* 248: pp. H186–H192, 1985.

[60] Cucciaro G, Yamaguchi Y, Mills E, Kuhn CM, Anthony DC, Branum GD, Epstein R, Meyers WC. Evaluation of selective liver denervation methods. *Am J Physiol* 259: pp. G781–G785, 1990.

[61] d'Almeida MS, Lautt WW. The effect of glucagon on vasoconstriction and vascular escape from nerve and noradrenaline-induced constriction of the hepatic artery of the cat. *Can J Physiol Pharmacol* 67: pp. 1418–1425, 1989.

[62] d'Almeida MS, Lautt WW. The effect of glucagon on autoregulatory escape from hepatic arterial vasoconstriction in the cat. *Proc West Pharmacol Soc* 32: pp. 265–267, 1989.

[63] d'Almeida MS, Lautt WW. Expression of vascular escape: conductance or resistance. *Am J Physiol* 262: pp. H1191–H1196, 1992.

[64] Dauzat M, Lafortune M, Patriquin H, Pomier-Layrargues G. Meal induced changes in hepatic and splanchnic circulation: a noninvasive Doppler study in normal humans. *Eur J Appl Physiol* 68: pp. 373–380, 1994.

[65] Deaciuc IV, D'Souza NB, Spitzer JJ. Alcohol-induced hepatic sinusoidal endothelial cell dysfunction in the mouse is mediated by Kupffer cells. *Int Hepatol Commun* 3: pp. 139–143, 1995.

[66] Decker K. Biologically active products of stimulated liver macrophages (Kupffer cells). *Eur J Biochem* 192: pp. 245–261, 1990.

[67] DeLeve LD, Wang X, Kuhlenkamp JF, Kaplowitz N. Toxicity of azathioprine and monocrotaline in murine sinusoidal endothelial cells and hepatocytes: the role of glutathione and relevance to hepatic venoocclusive disease. *Hepatology* 23: pp. 589–599, 1996.

[68] Dianzani MU. The role of free radicals in liver damage. *Proc Nutr Soc* 46: pp. 43–52, 1987.

[69] DiBona GF. Reflex regulation of renal sympathetic nerve activity in cirrhosis. In: *Liver and Nervous System. Falk Symposium No. 103*, edited by Häussinger D, Jungermann K. UK: Kluwer Academic Publishers, pp. 315–319, 1998.

[70] Di Luzio NR. A mechanism of the acute ethanol-induced fatty liver and the modification of liver injury by antioxidants. *Am J Pharm Sci Support Public Health* 15: pp. 50–63, 1966.

[71] Dive C, Nadalini AC, Heremans JF. Origin and composition of hepatic lymph proteins in the dog. *Lymphology* 4: pp. 133–139, 1971.

[72] Donald DE. Mobilization of blood from the splanchnic circulation. In: *Hepatic Circulation in Health and Disease*, edited by Lautt WW. New York: Raven Press, pp. 193–201, 1981.

[73] Dresel PE, Wallentin I. Effects of sympathetic vasoconstrictor fibres, noradrenaline and vasopressin on the intestinal vascular resistance during constant blood flow or blood pressure. *Acta Physiol Scand* 66: pp. 427–436, 1966.

[74] Eguchi S, Yanaga K, Sugiyama N, Okudaira S, Furui J, Kanematsu T. Relationship between portal venous flow and liver regeneration in patients after living donor right-lobe liver transplantation. *Liver Transpl* 9: pp. 547–551, 2003.

[75] Elias H, Petty D. Gross anatomy of the blood vessels and ducts within the human liver. *Am J Anat* 90: pp. 59–112, 1952.

[76] Ellis AJ, O'Grady JG. Clinical disorders of renal function in acute liver failure. In: *Ascites and Renal Dysfunction in Liver Disease: Pathogenesis, Diagnosis, and Treatment,* edited by Arroyo V, Ginès P, Rodés J, Schrier RW. Malden: Blackwell Science, pp. 36–62, 1999.

[77] Ezzat WR, Lautt WW. Hepatic arterial pressure-flow autoregulation is adenosine mediated. *Am J Physiol* 252: pp. H836–H845, 1987.

[78] Farrell GC, Zaluzny L. Portal vein ligation selectively lowers hepatic cytochrome P450 levels in rats. *Gastroenterology* 85: pp. 275–282, 1983.

[79] Fernandez-Munoz D, Caramelo C, Santos JC, Blanchart A, Hernando L, Lopez-Novoa JM. Systemic and splanchnic hemodynamic disturbances in conscious rats with experimental liver cirrhosis without ascites. *Am J Physiol* 249: pp. G316–G320, 1985.

[80] Folkow B. A critical study of some methods used in investigations on the blood circulation. *Acta Physiol Scand* 27: pp. 118–129, 1953.

[81] Folkow B, Lundgren O, Wallentin I. Studies on the relationship between flow resistance, capillary filtration coefficient and regional blood volume in the intestine of the cat. *Acta Physiol Scand* 57: pp. 270–283, 1963.

[82] Forssmann WF, Ito S. Hepatocyte innervations in primates. *J Cell Biol* 74: pp. 299–313, 1977.

[83] Francois-Franck CA, Hallion L. *Arch de Physiol* 8: pp. 908–922, 1896.

[84] Fraser R, Bowler LM, Day WA. Damage of rat liver sinusoidal endothelium by ethanol. *Pathology* 12: pp. 371–376, 1980.

[85] Fraser R, Day WA, Fernando NS. Atherosclerosis and the liver sieve. In: *Cells of the Hepatic Sinusoid Volume 1,* edited by Kirn A, Knook DL, Wisse E. Kupffer Cell Foundation, pp. 317–322, 1986.

[86] Fraser R, Dobbs BR, Rogers GWT. Lipoproteins and the liver sieve: the role of the fenestrated sinusoidal endothelium in lipoprotein metabolism, atherosclerosis, and cirrhosis. *Hepatology* 21: pp. 863–874, 1995.

[87] Gad J. Studies on the relations of the blood stream of the portal vein to the blood stream in the hepatic artery. *Dissertation,* G. Schade, Berlin, 1873.

[88] Garceau D, Yamaguchi N. Pharmacological evidence for the existence of a neuronal amine uptake mechanism in the dog liver in vivo. *Can J Physiol Pharmacol* 60: pp. 755–762, 1982.

[89] Garcia-Sainz JA, Hernandez-Munoz R, Santamaria A, Chagoya de Sanchez V. Mechanism of the fatty liver induced by cycloheximide and its reversibility by adenosine. *Biochem Pharmacol* 28: pp. 1409–1413, 1979.

[90] Gardemann A, Strulik H, Jungermann K. Nervous control of glycogenolysis and blood flow in arterially and portally perfused liver. *Am J Physiol* 253: pp. E238–E245, 1987.

[91] Gardner HW. Oxygen radical chemistry of polyunsaturated fatty acids. *Free Rad Biol Med* 7: pp. 65–86, 1989.

[92] Gelman S. General anesthesia and hepatic circulation. *Can J Physiol Pharmacol* 65: pp. 1762–1797, 1987.

[93] Gentilini P, Laffi G, Villa GL, Romanelli RG, Blendis LM. Ascites and hepatorenal syndrome during cirrhosis: two entities or the continuation of the same complication? *J Hepatology* 31: pp. 1088–1097, 1999.

[94] Gines P, Arroyo V. Hepatorenal syndrome. *J Am Soc Nephrol* 10: pp. 1833–1839, 1999.

[95] Gines P, Guevara M, Arroyo V, Rodes J. Hepatorenal syndrome. *Lancet* 362: pp. 1819–1827, 2003.

[96] Ginsburg M, Grayson J, Johnson DH. The nervous regulation of liver blood flow. *Proc Physiol Soc Lond* 17: 74P–75P, 1952.

[97] Goresky CA. Tracer behavior in the hepatic microcirculation. In: *Hepatic Circulation in Health and Disease*, edited by Lautt WW, New York: Raven Press, pp. 25–38, 1981.

[98] Goresky CA. Cell membrane transport processes: their role in hepatic uptake. In: *The Liver: Biology and Pathobiology*, edited by Aries I, Popper H, Schachter D, Shafritz DA. New York: Raven Press, pp. 581–599, 1982.

[99] Goresky CA, Cousineau D, Rose CP, Lee S. Lack of liver vascular response to carotid occlusion in mildly acidotic dogs. *Am J Physiol* 251: pp. H991–H999, 1986.

[100] Goresky CA, Groom AC. Microcirculatory events in the liver and spleen. In: *Handbook of Physiology—Cardiovascular System*, Volume IV. Bethesda, MD: Am. Physiol. Soc., pp. 689–780, 1986.

[101] Granger DN, Miller T, Allen R, Parker RE, Parker JC, Taylor AE. Permselectivity of cat liver blood–lymph barrier to endogenous macromolecules. *Gastroenterology* 77: pp. 103–109, 1979.

[102] Green HD, Hall LS, Sexton J, Deal CP. Autonomic vasomotor responses in the canine hepatic arterial and venous beds. *Am J Physiol* 196: pp. 196–202, 1959.

[103] Greenway CV. Hepatic plethysmography. In: *Hepatic Circulation in Health and Disease*, edited by Lautt WW. New York: Raven Press, pp. 41–56, 1981.

[104] Greenway CV. Mechanisms and quantitative assessment of drug effects on cardiac output using a new model of the circulation. *Pharmacol Rev* 33: pp. 213–251, 1981.

[105] Greenway CV. Hepatic fluid exchange. In: *Hepatic Circulation in Health and Disease*, edited by Lautt WW. New York: Raven Press, pp. 153–167, 1981.

[106] Greenway CV. Neural control and autoregulatory escape. In: *Physiology of the Intestinal Circulation*, edited by Shepherd AP, Granger DN. New York: Raven Press, pp. 61–71, 1984.

[107] Greenway CV. Autoregulatory escape in arteriolar resistance vessels. In: *Smooth Muscle Contraction*, edited by Stephens NL. New York: Marcel Dekker, pp. 473–484, 1984.

[108] Greenway CV. Effects of haemorrhage and hepatic nerve stimulation on venous compliance and unstressed volume in cat liver. *Can J Physiol Pharmacol* 65: pp. 2168–2174, 1987.

[109] Greenway CV, Burczynski F, Innes IR. Effects of bromocryptine on hepatic blood volume responses to hepatic nerve stimulation in cats. *Can J Physiol Pharmacol* 64: pp. 621–624, 1986.

[110] Greenway CV, Lautt WW. Effects of hepatic venous pressure on transsinusoidal fluid transfer in the liver of the anesthetized cat. *Circ Res* 26: pp. 697–703, 1970.

[111] Greenway CV, Lautt WW. Effects of adrenaline, isoprenaline and histamine on transsinusoidal fluid filtration in the cat liver. *Br J Pharmacol* 44: pp. 185–191, 1972.

[112] Greenway CV, Lautt WW. Blood volume, the venous system, preload, and cardiac output. *Can J Physiol Pharmacol* 64: pp. 383–387, 1986.

[113] Greenway CV, Lautt WW. Distensibility of hepatic venous resistance sites and consequences on portal pressure. *Am J Physiol* 254: pp. H452–H458, 1988.

[114] Greenway CV, Lautt WW. The hepatic circulation. In: *Handbook of Physiology—The Gastrointestinal System I*. Bethesda, MD: Am. Physiol. Soc., Volume 1, Part 2, Chapter 41, pp. 1519–1564, 1989.

[115] Greenway CV, Lautt WW. Acute and chronic ethanol on hepatic oxygen ethanol and lactate metabolism in cats. *Am J Physiol* 258: pp. G411–G418, 1990.

[116] Greenway CV, Lawson AE. β-Adrenergic receptors in the hepatic arterial bed of the anaesthetized cat. *Can J Physiol Pharmacol* 47: pp. 415–419, 1969.

[117] Greenway CV, Lawson AE, Mellander S. The effects of stimulation of the hepatic nerves, infusions of noradrenaline and occlusion of the carotid arteries on liver blood flow in the anesthetized cat. *J Physiol* 192: pp. 21–41, 1967.

[118] Greenway CV, Lawson AE, Stark RD. Vascular responses of the spleen to nerve stimulation during normal and reduced blood flow. *J Physiol Lond* 194: pp. 421–433, 1968.

[119] Greenway CV, Oshiro G. Intrahepatic distribution of portal and hepatic arterial blood flows in anaesethetized cats and dogs and the effects of portal occlusion, raised venous pressure and histamine. *J Physiol* 227: pp. 473–485, 1972.

[120] Greenway CV, Oshiro G. Comparison of the effects of hepatic nerve stimulation on arterial flow, distribution of arterial and portal flows and blood content in the livers of anaesthetized cats and dogs. *J Physiol* 227: pp. 487–501, 1972.

[121] Greenway CV, Seaman KL, Innes IR. Norepinephrine on venous compliance and unstressed volume in cat liver. *Am J Physiol* 248: pp. H468–H476, 1985.

[122] Greenway CV, Stark RD. Hepatic vascular bed. *Physiol Rev* 51: pp. 23–65, 1971.

[123] Greenway CV, Stark RD, Lautt WW. Capacitance responses and fluid exchange in the cat liver during stimulation of the hepatic nerves. *Circ Res* 25: pp. 277–284, 1969.

[124] Grisham JW, Nopanitaya W. Scanning electron microscopy of casts of hepatic microvessels: review of methods and results. In: *Hepatic Circulation in Health and Disease*, edited by Lautt WW. New York: Raven Press, pp. 87–109, 1981.

[125] Groszmann RJ, Blei AT, Kniaz JL, Storer EH, Conn HO. Portal pressure reduction induced by partial mechanical obstruction of the superior mesenteric artery in the anesthetized dog. *Gastroenterology* 75: pp. 187–192, 1978.

[126] Gugler R, Lain P, Azarnoff DL. Effect of portacaval shunt on the disposition of drugs with and without first-pass effect. *J Pharmacol Exp Ther* 195: pp. 416–423, 1975.

[127] Gulberg V, Haag K, Rossle M, Gerbes AL. Hepatic arterial buffer response in patients with advanced cirrhosis. *Hepatology* 35: pp. 630–634, 2002.

[128] Gumucio DL. Functional and anatomic heterogeneity in the liver acinus: impact on transport. *Am J Physiol* 244: pp. G578–G582, 1983.

[129] Heikkinen E, Larmi T. Immediate effect of partial hepatectomy on portal pressure in rats. *Acta Chir Scand* 134: pp. 367–368, 1968.

[130] Helyar L, Bundschuh DS, Laskin JD, Laskin DL. Induction of hepatic Ito cell nitric oxide production after acute endotoxemia. *Hepatology* 20: pp. 1509–1515, 1994.

[131] Henderson JM, Gilmore GT, Mackay GJ, Galloway JR, Dodson TF, Kutner MH. Hemodynamics during liver transplantation: the interactions between cardiac output and portal venous and hepatic arterial flows. *Hepatology* 16: pp. 715–718, 1992.

[132] Henriksen JH, Ring-Larsen H, Christensen NJ. Autonomic nervous function in liver disease. In: *Cardiovascular Complications in Liver Disease*, edited by Bomzon A, Blendis LM. Boston: CRC Press, pp. 63–79, 1990.

[133] Hernandez-Munoz R, Santamaria A, Garcia-Sainz JA, Pina E, Chagoya de Sanchez V. On the mechanism of ethanol-induced fatty liver and its reversibility by adenosine. *Arch Biochem Biophys* 190: pp. 155–162, 1978.

[134] Hickey PL, Angus PW, McLean AJ, Morgan DJ. Oxygen supplementation restores theophylline clearance to normal in cirrhotic rats. *Gastroenterology* 108: pp. 1504–1509, 1995.

[135] Hinshaw LB, Vick JA, Jordan MM, Wittmers LE. Vascular changes associated with development of irreversible endotoxin shock. *Am J Physiol* 202: pp. 103–110, 1962.

[136] Hirata K, Ogata I, Ohta Y, Fujiwara K. Hepatic sinusoidal cell destruction in the development of intravascular coagulation in acute liver failure of rats. *J Pathol* 158: pp. 157–165, 1989.

[137] Holmin T, Ekelund M, Kullendorff C-M, Lindfeldt J. A mirosurgical method for denervation of the liver in the rat. *Eur J Surg Res* 16: pp. 288–293, 1984.

[138] Hortelano S, Dewez B, Genaro AM, Diaz-Guerra MJM, Bosca L. Nitric oxide is released in regenerating liver after partial hepatectomy. *Hepatology* 21: pp. 776–786, 1995.

[139] Hughes RL, Mathie RT, Fitch W, Campbell D. Liver blood flow and oxygen consumption during metabolic acidosis and alkalosis in the greyhound. *Clin Sci* 60: pp. 355–361, 1980.

[140] Hsu C-T. The role of the sympathetic nervous system in promoting liver cirrhosis induced by carbon tetrachloride, using the essential hypertensive animal (SHR). *J Auton Nerv Syst* 37: pp. 163–174, 1992.

[141] Hyatt RE, Lawrence GH, and Smith JR. Observations on the origin of ascites from experimental hepatic congestion. *J Lab Clin Med* 45: pp. 274–280, 1955.

[142] Ingles AC, Legare DJ, Lautt WW. Development of portacaval shunts in portal-stenotic cats. *Can J Physiol Pharmacol* 71: pp. 671–674, 1993.

[143] Ingles AC, Legare DJ, Lautt WW. Evaluation of vascular tone in portacaval shunts comparing the index of contractility and resistance in cats. *Hepatology* 20: pp. 1242–1246, 1994.

[144] Ishii K, Shimizu M, Karube H, Shibuya A, Shibata H, Okudaira M, Nagata H, Tsuchiya M. Inhibitory effect of noradrenaline on acute liver injury induced by carbon tetrachloride in the rat. *J Auton Nerv Syst* 39: pp. 13–18, 1992.

[145] Israel Y, Kalant H, Orrego H, Khanna JM, Videla L, Phillips JM. Experimental alcohol-induced hepatic necrosis: suppression by propylthiouracil. *Proc Natl Acad Sci USA* 72: pp. 1137–1141, 1975.

[146] Iwai M, Gardemann A, Puschel G, Jungermann K. Potential role for prostaglandin $F_{2\alpha}$, D_2, E_2 and thromboxane A_2 in mediating the metabolic and hemodynamic actions of sympathetic nerves in perfused rat liver. *Eur J Biochem* 175: pp. 45–50, 1988.

[147] Iwai M, Jungermann K. Possible involvement of eicosanoids in the actions of sympathetic hepatic nerves on carbohydrate metabolism and hemodynamics in perfused rat liver. *FEBS Lett* 221: pp. 155–160, 1987.

[148] Iwai M, Jungermann K. Mechanism of action of cysteinyl leukotrienes on glucose and lactate balance and on flow in perfused rat liver. *Eur J Biochem* 180: pp. 273–281, 1989.

[149] Iwai M, Saheki S, Ohta Y, Shimazu T. Footshock stress accelerates carbon tetrachloride-induced liver injury in rats: implication of the sympathetic nervous system. *Biomed Res* 7: pp. 145–154, 1986.

[150] Iwai M, Shimazu T. Exaggeration of acute liver damage by hepatic sympathetic nerves and circulating catecholamines in perfused liver of rats treated with D-galactosamine. *Hepatology* 23: pp. 524–529, 1996.

[151] Iwai M, Shimazu T. Effects of ventromedial and lateral hypothalamic stimulation on chemically-induced liver injury in rats. *Life Sci* 42: pp. 1833–1840, 1988.

[152] Iwao T, Toyonaga A, Shigemori H, Oho K, Sakai T, Tayama C, Masumoto H, Sato M, Tanikawa K. Hepatic artery hemodynamic responsiveness to altered portal blood flow in normal and cirrhotic livers. *Radiology* 200: pp. 793–798, 1996.

[153] Jalan R, Forrest EH, Redhead DN, Dillon JF, Hayes PC. Reduction in renal blood flow following acute increases in the portal pressure: evidence for the existence of a hepatorenal reflex in man? *Gut* 40: pp. 664–670, 1997.

[154] Jarhult J. Osmotic fluid transfer from tissue to blood during hemorrhagic hypotension. *Acta Physiol Scand* 89: pp. 213–226, 1973.

[155] Ji S, Beckh K, Jungermann K. Regulation of oxygen consumption and microcirculation by α-sympathetic nerves in isolated perfused rat liver. *FEBS Lett* 167: pp. 117–122, 1984.

[156] Johnson PC. Autoregulation of intestinal blood flow. Am J Physiol 199: pp. 311–318, 1960.

[157] Jungermann K, Katz N. Functional specialization of different hepatocyte populations. *Physiol Rev* 69: pp. 708–764, 1989.

[158] Jungermann K, Kietzmann T. Zonation of parenchymal and nonparenchymal metabolism in liver. *Ann Rev Nutr* 16: pp. 179–203, 1996.

[159] Kamiya A, Togawa T. Adaptive regulation of wall shear stress to flow change in the canine carotid artery. *Am J Physiol* 239: pp. H14–H21, 1980.

[160] Kawada N, Tran-Thi T, Klein H, Decker K. The contraction of hepatic stellate (Ito) cells stimulated with vasoactive substance. Possible involvement of endothelin 1 and nitric oxide in the regulation of the sinusoidal tonus. *Eur J Biochem* 213: pp. 815–823, 1993.

[161] Kawada N, Klein H, Decker K. Eicosanoid-mediated contractility of hepatic stellate cells. *Biochem J* 285: pp. 367–371, 1992.

[162] Kawada N, Kuroki T, Uoya M, Inoue M, Kobayashi K. Smooth muscle α-actin expression in rat hepatic stellate cell is regulated by nitric oxide and cGMP production. *Biochem Biophys Res Commun* 229: pp. 238–242, 1996.

[163] Kawasaki T, Moriyasu F, Kimura T, Someda H, Tamada T, Yamashita Y, Ono S, Kajimura K, Hamato N, Okuma M. Hepatic function and portal hemodynamics in patients with liver cirrhosis. *Am J Gastroenterol* 85: pp. 1160–1164, 1990.

[164] Kehrer JP. Free radicals as mediators of tissue injury and disease. *Clin Rev Toxicol* 23: pp. 21–48, 1993.

[165] Keiding S. Drug administration to liver patients: aspects of liver pathophysiology. *Sem Liver Dis* 15: pp. 268–282, 1995.

[166] Kelm M, Feelisch M, Seussen A, Strauer BE, Schrader J. Release of endothelium derived nitric oxide in relation to pressure and flow. *Cardiovasc Res* 25: pp. 831–836, 1991.

[167] Kennedy C, Burnstock G. Evidence for an inhibitory prejunctional P1-purinoceptor in the rat portal vein with characteristics of the A_2 rather than of the A_1 subtype. *Eur J Pharmacol* 100: pp. 363–368, 1984.

[168] Koepke JP, Jones S, DiBona GF. Renal nerves mediate blunted natriuresis to atrial natriuretic peptide in cirrhotic rats. *Am J Physiol* 252: pp. R1019–R1023, 1987.

[169] Konovalova GG, Lankin VZ, Beskrovnova NN. The role of free radical inhibitors of lipid peroxidation in protecting the myocardium from ischemic damage. *Arkh Patol* 51: pp. 19–24, 1989.

[170] Koo A, Liang IYS. Parasympathetic cholinergic vasodilator mechanism in the terminal liver microcirculation in rats. *Quart J Exp Physiol* 64: pp. 149–159, 1979.

[171] Koyoma S, Kanai K, Aibiki M, Fujita T. Reflex increase in renal nerve activity during acutely altered portal venous pressure. *J Auton Nerv Syst* 23: pp. 55–62, 1988.

[172] Krarup N. The effects of noradrenaline and adrenaline on hepatosplanchnic hemodynamics, functional capacity of the liver and hepatic metabolism. *Acta Physiol Scand* 87: pp. 307–319, 1973.

[173] Kukielka E, Dicker E, Cederbaum AI. Increased production of reactive oxygen species by rat liver mitochondria after chronic ethanol treatment. *Arch Biochem Biophys* 309: pp. 377–386, 1994.

[174] Kuiper J, Brouwer A, Knook DL, van Berkel TJC. Kupffer and sinusoidal endothelial cells. In: *The Liver Biology and Pathobiology*, Third Edition, edited by Arias IM, Boyer JL, Fausto N, Jakoby WB, Schachter DA, Shafritz DA. New York: Raven Press, Chapter 41, pp. 791–818, 1994.

[175] Kurose I, Kato S, Ishii H, Fukumura D, Miura S, Suematsu M, Tsuchiya M. Nitric oxide mediates lipopolysaccharide-induced alteration of mitochondrial function in cultures hepatocytes and isolated perfused liver. *Hepatology* 18: pp. 380–388, 1993.

[176] Kurose I, Miura S, Fukumura D, Yonei Y, Saito H, Tada S, Suematsu M, Tsuchiya M. Nitric oxide mediated Kupffer cell-induced reduction of mitochondrial energization in hepatoma cells: a comparison with oxidative burst. *Cancer Res* 53: pp. 2676–2682, 1993.

[177] Kurose I, Miura S, Higuchi H, Watanabe N, Kamegaya Y, Takaishi M, Tomita K, Fukumura D, Kato S, Ishii H. Increased nitric oxide synthase activity as a cause of mitochon-

drial dysfunction in rat hepatocytes: roles for tumor necrosis factor alpha. *Hepatology* 24: pp. 1185–1192, 1996.

[178] Kyosola K, Pentitila O, Ihamaki T, Varis K, Salaspuro M. Adrenergic innervations of the human liver. *Scand J Gastroenterol* 20: pp. 254–256, 1985.

[179] Lafortune M, Dauzat M, Pomier-Layrargues G, Gianfelice D, Lepanto L, Breton G, Marleau D, Dagenais M, Lapointe R. Hepatic artery: effect of a meal in healthy persons and transplant recipients. *Radiology* 187: pp. 391–394, 1993.

[180] Laine G, Hall JT, Laine SH, and Granger HJ. Transsinusoidal fluid dynamics in canine liver during venous hypertension. *Circ Res* 45: pp. 317–323, 1979.

[181] Lamers WH, Hoynes KE, Zonneveld D, Moorman AFM, Charles R. Noradrenergic innervation of developing rat and spiny mouse liver: its relation to the development of the liver architecture and enzymic zonation. *Anat Embryol* 178: pp. 175–181, 1988.

[182] Lang F, Tschernko E, Schulze E, Ottl I, Ritter M, Volkl H, Hallbrucker C, Haussinger D. Hepatorenal reflex regulating kidney function. *Hepatology* 14: pp. 590–594, 1991.

[183] Latour MG, Lautt WW. The hepatic vagus nerve in the control of insulin sensitivity in the rat. *Auton Neurosci* 95: pp. 125–130, 2002.

[184] Lautt WW. Method for measuring hepatic uptake of oxygen or other blood-borne substances in situ. *J Appl Physiol* 40: pp. 269–274, 1976.

[185] Lautt WW. Control of hepatic and intestinal blood flow: effect of isovolemic haemodilution on blood flow and oxygen uptake in the intact liver and intestines. *J Physiol Lond* 265: pp. 313–326, 1977.

[186] Lautt WW. The hepatic artery: subservient to hepatic metabolism or guardian of normal hepatic clearance rates of humoral substance. *Gen Pharmacol* 8: pp. 73–78, 1977.

[187] Lautt WW. Effect of stimulation on hepatic nerves on hepatic O_2 uptake and blood flow. *Am J Physiol* 232: pp. H652–H656, 1977.

[188] Lautt WW. Hepatic presinusoidal sphincters affected by altered arterial pressure and flow, venous pressure and nerve stimulation. *Microvasc Res* 15: pp. 309–317, 1978.

[189] Lautt WW. Neural activation of α-adrenoreceptors in glucose mobilization from liver. *Can J Physiol Pharmacol* 57: pp. 1037–1039, 1979.

[190] Lautt WW. Hepatic nerves—a review of their functions and effects. *Can J Physiol Pharmacol* 58: pp. 105–123, 1980.

[191] Lautt WW. Control of hepatic arterial blood flow: independent from liver metabolic activity. *Am J Physiol* 239: pp. H559–H564, 1980.

[192] Lautt WW. Role and control of the hepatic artery. In: *Hepatic Circulation in Health and Disease*, edited by Lautt WW. New York: Raven Press, pp. 203–226, 1981.

[193] Lautt WW. Evaluation of surgical denervation of the liver in cats. *Can J Physiol Pharmacol* 59: pp. 1013–1016, 1981.

[194] Lautt WW. Carotid sinus baroreceptors effects on cat livers in control and haemorrhaged state. *Can J Physiol Pharmacol* 60: pp. 1592–1602, 1982.

[195] Lautt WW. Relationship between hepatic blood flow and overall metabolism: the hepatic arterial buffer response. *Fed Proc* 42: pp. 1662–1666, 1983.

[196] Lautt WW. Afferent and efferent neural roles in liver function. *Prog Neurobiol* 21: pp. 323–348, 1983.

[197] Lautt WW. Autoregulation of superior mesenteric artery is blocked by adenosine antagonism. *Can J Physiol Pharmacol* 64: pp. 1291–1295, 1986.

[198] Lautt WW. Resistance or conductance for expression of arterial vascular tone. *Microvasc Res* 37: pp. 230–236, 1989.

[199] Lautt WW. Non-competitive antagonism of adenosine by caffeine on the hepatic and superior mesenteric arteries of anesthetized cats. *J Pharmacol Exp Ther* 254: pp. 400–406, 1990.

[200] Lautt WW. The HISS story overview: a novel hepatic neurohumoral regulation of peripheral insulin sensitivity in health and diabetes. *Can J Physiol Pharmacol* 77: pp. 553–562, 1999.

[201] Lautt WW. Practice and principles of pharmacodynamic determination of HISS-dependent and HISS-independent insulin action: methods to quantitate mechanisms of insulin resistance. *Med Res Rev* 23: pp. 1–14, 2003.

[202] Lautt WW. A new paradigm for diabetes and obesity: the hepatic insulin sensitizing substance (HISS) hypothesis. *J Pharmacol Sci* 95: pp. 9–17, 2004.

[203] Lautt WW. Postprandial insulin resistance as an early predictive cardiovascular risk factor. *Therap Clin Risk Managem* 3: pp. 761–770, 2007.

[204] Lautt WW, Brown LC, Durham JS. Active and passive control of hepatic blood volume responses to hemorrhage at normal and raised hepatic venous pressure in cats. *Can J Physiol Pharmacol* 58: pp. 1049–1057, 1980.

[205] Lautt WW, Carroll AM. Evaluation of topical phenol as a means of producing autonomic denervation of the liver. *Can J Physiol Pharmacol* 62: pp. 849–853, 1984.

[206] Lautt WW, Cote MG. Functional evaluation of 6-hydroxydopamine-induced sympathectomy in the liver of the cat. *J Pharmacol Exp Ther* 198: pp. 562–567, 1976.

[207] Lautt WW, Cote MG. The effect of 6-hydroxydopamine-induced hepatic sympathectomy in the liver of the cat. J Trauma 17: pp. 270–274, 1977.

[208] Lautt WW, d'Almeida MS, McQuaker J, D'ALeo L. Impact of the hepatic arterial buffer response on splanchnic vascular responses to intravenous adenosine, isoproterenol and glucagon. *Can J Physiol Pharmacol* 66: pp. 807–813, 1988.

[209] Lautt WW, Daniels TR. Differential effect of taurocholic acid on hepatic arterial resistance vessels and bile flow. *Am J Physiol* 244: pp. G366–G369, 1983.

[210] Lautt WW, Dwan PD, Singh RR. Control of the hyperglycemic response to hemorrhage in cats. *Can J Physiol Pharmacol* 60: pp. 1618–1623, 1982.

[211] Lautt WW, Greenway CV. Hepatic capacitance vessel responses to bilateral carotid occlusion in anesthetized cats. *Can J Physiol Pharmacol* 50: pp. 244–247, 1972.

[212] Lautt WW, Greenway CV. Hepatic venous compliance and role of liver as a blood reservoir. *Am J Physiol* 231: pp. 292–295, 1976.

[213] Lautt WW, Greenway CV. Conceptual review of the hepatic vascular bed. *Hepatology* 7: pp. 952–963, 1987.

[214] Lautt WW, Greenway CV, Legare DJ. Index of contractility: quantitative analysis of hepatic venous distensibility. *Am J Physiol* 260: pp. G325–G332, 1991.

[215] Lautt WW, Greenway CV, Legare DJ, Weisman H. Localization of intrahepatic portal vascular resistance. *Am J Physiol* 251: pp. G375–G381, 1986.

[216] Lautt WW, Legare DJ. The use of 8-phenyltheophylline as a competitive antagonist of adenosine and an inhibitor of the intrinsic regulatory mechanism of the hepatic artery. *Can J Physiol Pharmacol* 63: pp. 717–722, 1985.

[217] Lautt WW, Legare DJ. Adenosine modulation of hepatic arterial but not portal venous constriction induced by sympathetic nerves, norepinephrine, angiotensin, and vasopressin in the cat. *Can J Physiol Pharmacol* 64: pp. 449–454, 1986.

[218] Lautt WW, Legare DJ. Evaluation of hepatic venous resistance responses using index of contractility (IC). *Am J Physiol* 262: pp. G510–G516, 1992.

[219] Lautt WW, Legare DJ. Passive autoregulation of portal venous pressure: distensible hepatic resistance. *Am J Physiol* 263: pp. G702–G708, 1992.

[220] Lautt WW, Legare DJ, d'Almeida MS. Adenosine as putative regulator of hepatic arterial flow (the buffer response). *Am J Physiol* 248: pp. H331–H338, 1985.

[221] Lautt WW, Legare DJ, Daniels TR. The comparative effect of administration of substances via the hepatic artery or portal vein on hepatic arterial resistance, liver blood volume and hepatic extraction in cats. *Hepatology* 4: pp. 927–932, 1984.

[222] Lautt WW, Legare DJ, Ezzat WR. Quantitation of the hepatic arterial buffer response to graded changes in portal blood flow. *Gastroenterology* 98: pp. 1024–1028, 1990.

[223] Lautt WW, Legare DJ, Greenway CV. Effect of hepatic venous sphincter contraction on transmission of central venous pressure to lobar and portal pressure. *Can J Physiol Pharmacol* 65: pp. 2235–2243, 1987.

[224] Lautt WW, Legare DJ, Turner GA. Evaluation of hepatic venous balloon occluder to estimate portal pressure. *Clin Invest Med* 13: pp. 247–255, 1990.

[225] Lautt WW, Lockhart LK, Legare DJ. Adenosine modulation of vasoconstrictor responses to sympathetic nerves and noradrenaline infusion in the superior mesenteric artery of the cat. *Can J Physiol Pharmacol* 66: pp. 937–941, 1988.

[226] Lautt WW, Macedo MP. Nitric oxide and hepatic circulation. In: *Nitric Oxide and the Regulation of the Peripheral Circulation*, edited by Kadowitz PJ, McNamara DB. Boston: Birkhauser, pp. 243–258, 2000.

[227] Lautt WW, McQuaker JE. Maintenance of hepatic arterial blood flow during hemorrhage is mediated by adenosine. *Can J Physiol Pharmacol* 67: pp. 1023–1028, 1989.

[228] Lautt WW, Ming Z, Legare DJ. Attenuation of age- and sucrose-induced insulin resistance and syndrome X by a synergistic antioxidant cocktail. Canadian Oxidative Stress Consortium Paper. *Can J Physiol Pharmacol* (submitted for publication), 2009.

[229] Lautt WW, Ming Z, Macedo MP, Legare DJ. HISS-dependent insulin resistance (HDIR) in aged rats is associated with adiposity, progresses to syndrome X, and is attenuated by a unique antioxidant cocktail. *Exp Gerontol* 43: pp. 790–800, 2008.

[230] Lautt WW, Plaa GL. Hemodynamic effects of CCL4 in the intact liver of the cat. *Can J Physiol Pharmacol* 52: pp. 727–735, 1974.

[231] Lautt WW, Schafer J, Legare DJ. Effect of adenosine and glucagon on hepatic blood volume responses to sympathetic nerves. *Can J Physiol Pharmacol* 69: pp. 43–48, 1991.

[232] Lautt WW, Schafer J, Legare DJ. Hepatic blood flow distribution: consideration of gravity, liver surface, and norepinephrine on regional heterogeneity. *Can J Physiol Pharmacol* 71: pp. 128–135, 1993.

[233] Lautt WW, Schafer J, Macedo MP, Legare DJ. Repurposing two pharmaceuticals, bethanechol and *N*-acetylcysteine, to mimic both essential feeding signals that regulate meal-induced insulin sensitization, and to reverse insulin resistance in fasted and sucrose-induced diabetic rats. *Submitted for publication*, 2010.

[234] Lautt WW, Skelton FS. Effect of hepatic nerve stimulation on hepatic uptake of lidocaine in the cat. *Life Sci* 19: pp. 433–436, 1976.

[235] Lautt WW, Wong C. Hepatic parasympathetic neural effect on glucose balance in the intact liver. *Can J Physiol Pharmacol* 56: pp. 679–682, 1978.

[236] Legare DJ, Lautt WW. Hepatic venous resistance site in the dog: localization and validation of intrahepatic pressure measurements. *Can J Physiol Pharmacol* 65: pp. 352–359, 1987.

[237] Lerner MH, Lowy BA. The formation of adenosine in rabbit liver and its possible role as a direct precursor of erythrocyte adenine nucleotides. *J Biol Chem* 249: pp. 959–966, 1974.

[238] Levy M, Wexler MJ. Renal sodium retention and ascites formation in dogs with experimental cirrhosis but without portal hypertension or increased splanchnic vascular capacity. *Lab Clin Med* 91: pp. 520–536, 1978.

[239] Liang CC. The influence of hepatic portal circulation on urine flow. *J Physiol* 214: pp. 571–581, 1977.

[240] Lieber CS. Mechanism of ethanol induced hepatic injury. *Pharmac Ther* 46: pp. 1–41, 1990.

[241] Lind J. Changes in the liver circulation at birth. *Ann N Y Acad Sci* 111: pp. 110–120, 1963.

[242] Liu P, McGuire GM, Fisher MA, Farhood A, Smith CW, Jaeschke J. Activation of Kupffer cells and neutrophils for reactive oxygen formation is responsible for endotoxin-enhanced liver injury after hepatic ischemia. Shock 3: pp. 56–62, 1995.

[243] Lloyd HE, Schrader J. The importance of the transmethylation pathway for adenosine metabolism in the heart. In: *Topics and Perspectives in Adenosine Research*, edited by Gerlach E, Becker BF. Berlin: Springer-Verlag, 1987.

[244] Lockhart LK, Legare DJ, Lautt WW. Kinetics of adenosine antagonism of sympathetic nerve-induced vasoconstriction. *Proc West Pharmacol Soc* 31: pp. 105–107, 1988.

[245] Lohse CL, Suter PF. Functional closure of the ductus venosus during early postnatal life in the dog. *Am J Vet Res* 38: pp. 839–844, 1977.

[246] Lough J, Rosenthall L, Arzoumanian A, Goresky CA. Kupffer cell depletion associated with capillarization of liver sinusoids in carbon tetrachloride-induced rat liver cirrhosis. *J Hepatology* 5: pp. 190–198, 1987.

[247] Louis-Sylvester J, Servant JM, Molimard R, Le Magnen J. Effect of liver denervation on the feeding pattern of fats. *Am J Physiol* 239: pp. R66–R70, 1980.

[248] Maass-Moreno R, Rothe CF. Carotid baroreceptors control of liver and spleen volume in cats. *Am J Physiol* 260: pp. H254–H259, 1991.

[249] Macedo MP, Lautt WW. Nitric oxide synthase antagonism potentiates pressure-flow autoregulation in the superior mesenteric artery. *Proc West Pharmacol Soc* 38: pp. 33–34, 1995.

[250] Macedo MP, Lautt WW. Shear-induced modulation by nitric oxide of sympathetic nerves in the superior mesenteric artery. *Can J Physiol Pharmacol* 74: pp. 692–700, 1996.

[251] Macedo MP, Lautt WW. Potentiation to vasodilators by nitric oxide synthase blockade in superior mesenteric but not hepatic artery. *Am J Physiol* 270: pp. G507–G514, 1997.

[252] Macedo MP, Lautt WW. Shear-induced modulation of vasoconstriction in the hepatic artery and portal vein by nitric oxide. *Am J Physiol* 274: pp. G253–G260, 1998.

[253] MacPhee PJ, Schmidt EE, Groom AC. Evidence for Kupffer cell migration along liver sinusoids, from high resolution in vivo microscopy. *Am J Physiol* 263: pp. G17–G23, 1992.

[254] Mathie RT, Alexander B. The role of adenosine in hyperaemic response of the hepatic artery to portal vein occlusion (the 'buffer response'). *Br J Pharmacol* 100: pp. 626–630, 1990.

[255] Mathie RT, Blumgart LH. The hepatic hemodynamic response to acute portal venous blood flow reductions in the dog. *Pflugers Arch* 399: pp. 223–227, 1983.

[256] Mathie RT, Blumgart LH. Effect of denervation on the hepatic hemodynamic response to hypercapnia and hypoxia in the dog. *Pflugers Arch* 397: pp. 152–157, 1983.

[257] Mathie RT, Lam PHM, Harper AM, Blumgart LH. The hepatic arterial blood flow response to portal vein occlusion in the dog. *Pflugers Arch* 386: pp. 77–83, 1980.

[258] Matsumura T, Thurman RG. Predominance of glycolysis in pericentral regions of liver lobule. *Eur J Biochem* 140: pp. 229–234, 1984.

[259] McCord JM. Oxygen-derived free radicals in postischemic tissue injury. *N Engl J Med* 312: pp. 159–163, 1985.

[260] McCuskey RS, Urbaschek R, McCuskey PA, Urbaschek B. In vivo microscopic studies of the responses of the liver to endotoxin. *Klin Wochenschr* 60: pp. 749–751, 1982.

[261] McKay T, Zink J, Greenway CV. Relative rates of absorption of fluid and protein from the peritoneal cavity in cats. *Lymphology* 11: pp. 106–110, 1978.

[262] McLain GE, Sipes IG, Brown BB. An animal model of halothane hepatotoxicity. *Anesthesiology* 51: pp. 321–326, 1979.

[263] Mehendale HM, Roth RA, Gandolfi AJ, Klaunig JE, Lemasters JJ, Curtis LR. Novel mechanisms in chemically induced hepatoxicity. *FASEB J* 8: pp. 1285–1295, 1994.

[264] Mellander S, Johansson B. Control of resistance, exchange, and capacitance functions in the peripheral circulation. *Pharmacol Rev* 20: pp. 117–196, 1968.

[265] Messerli FH, Nowaczynski W, Honda M, Genest J, Boucher R, Kuchel O, Rojoorte JM. Effects of angiotensin II on steroid metabolism and hepatic blood flow in man. *Circ Res* 40: pp. 204–207, 1977.

[266] Metz W, Forssmann WG. Innervation of the liver in guinea-pig and rat. *Anat Embryol* 160: pp. 239–252, 1980.

[267] Michalopolous GK, DeFrances MC. Liver regeneration. *Science* 276: pp. 60–66, 1997.

[268] Miller DL. Quantitative morphological assessment of the sinusoids of the hepatic acinus. In: *Hepatic Circulation in Health and Disease,* edited by Lautt WW. New York: Raven Press, pp. 111–135, 1981.

[269] Ming Z, Fan YJ, Yang X, Lautt WW. Blockade of intrahepatic adenosine receptors improves urine excretion in cirrhotic rats induced by thioacetamide. *J Hepatol* 42: pp. 680–686, 2005.

[270] Ming Z, Fan YJ, Yang X, Lautt WW. Contribution of hepatic adenosine A_1 receptors to renal dysfunction associated with acute liver injury in rats. *Hepatology* 44: pp. 813–822, 2006.

[271] Ming Z, Han C, Lautt WW. Nitric oxide mediates hepatic arterial vascular escape from norepinephrine-induced constriction. *Am J Physiol* 277: pp. G1200–G1206, 1999.

[272] Ming Z, Han C, Lautt WW. Nitric oxide inhibits norepinephrine-induced hepatic vascular responses but potentiates hepatic glucose output. *Can J Physiol Pharmacol* 78: pp. 36–44, 2000.

[273] Ming Z, Legare DJ, Lautt WW. Obesity, syndrome X, and diabetes: the role of HISS-dependent insulin resistance altered by sucrose, an antioxidant cocktail, and age. *Can J Physiol Pharmacol* (in press), 2009.

[274] Ming Z, Smyth DD, Lautt WW. Intrahepatic adenosine triggers a hepatorenal reflex to regulate sodium and water excretion. *Auton Neurosci* 93: pp. 1–7, 2001.

[275] Ming Z, Smyth DD, Lautt WW. Decreases in portal flow trigger a hepatorenal reflex to inhibit renal sodium and water excretion in rats: role of adenosine. *Hepatology* 35: pp. 167–175, 2002.

[276] Montano N, Lombardi F, Ruscone TG, Contini M, Guazzi M, Malliani A. The excitatory effect of adenosine on the discharge activity of the afferent cardiac sympathetic fibers. *Cardiologia* 36: pp. 953–959, 1991.

[277] Moreau R, Lee SS, Hadengue A, Braillon A, Lebrec D. Hemodynamic effects of a clonidine-induced decrease in sympathetic tone in patients with cirrhosis. *Hepatology* 7: pp. 149–154, 1987.

[278] Mori T, Okanoue T, Kanaoka H, Sawa Y, Kashima K. Experimental study of the reversibility of sinusoidal capillarization. *Alcohol Alcoholism* 29: pp. 67–74, 1994.

[279] Mosca P, Fa-Yauh L, Kaumann AJ, Groszmann RJ. Pharmacology of portal-systemic collaterals in portal hypertensive rats: role of endothelium. *Am J Physiol* 263: pp. G544–G550, 1992.

[280] Moser MJ, Gong Y, Zhang MN, Johnston J, Lipschitz J, Kneteman NM, Minuk GY. Immediate-early oncogene expression and liver function following varying extents of partial hepatectomy in the rat. *Dig Dis Sci* 46: pp. 907–914, 2001.

[281] Moselhy SS, Ali HK. Hepatoprotective effect of cinnamon extracts against carbon tetrachloride induced oxidative stress and liver injury in rats. *Biol Res* 42: pp. 93–98, 2009.

[282] Mucke I, Richter S, Menger MD, Vollmar B. Significance of hepatic arterial responsiveness for adequate tissue oxygenation upon portal vein occlusion in cirrhotic livers. *Int J Colorectal Dis* 15: pp. 335–341, 2000.

[283] Muller E, Colombo JP, Peheim E, Bircher J. Histochemical demonstration of gamma-glutamyltranspeptidase in rat liver after portacaval anastomosis. *Experientia* 30: pp. 1128–1129, 1974.

[284] Mundschau GA, Zimmerman SW, Gildersleeve JW, Murphy QR. Hepatic and mesenteric artery resistances after sinoaortic denervation and haemorrhage. *Am J Physiol* 211: pp. 77–82, 1966.

[285] Myers JD, Brannon ES, Holland BC. A correlative study of the cardiac output and the hepatic circulation in hyperthyroidism. *J Clin Invest* 29: pp. 1069–1077, 1950.

[286] Nieminen A-L, Dawson TL, Gores GJ, Kawanishi T, Herman B, Lemasters JJ. Protection by acidotic pH and fructose against lethal injury to rat hepatocytes from mitochondrial

inhibitors, inophores and oxidant chemicals. *Biochem Biophys Res Commun* 167: pp. 600–606, 1990.

[287] Nies AS, Wilkinson GR, Rush BD, Strother JT, McDevitt DG. Effects of alteration of hepatic microsomal enzyme activity on liver blood flow in the rat. *Biochem Pharmacol* 25: pp. 1991–1993, 1976.

[288] Niijima A. An electrophysiological study on hepatovisceral reflex: the role played by vagal hepatic afferents from chemosensors in the hepatoportal region. In: *Liver and Nervous System*, edited by Haussinger D, Jungermann K. Great Britain: Kluwer Academic Publishers, pp. 159–172, 1998.

[289] Nishida J, McCuskey RS, McDonnell D, Fox ES. Protective role of NO in hepatic microcirculatory dysfunction during endotoxemia. *Am J Physiol* 267: pp. G1135–G1141, 1994.

[290] O'Beirne JP, Chouhan M, Hughes RD. The role of infection and inflammation in the pathogenesis of hepatic encephalopathy and cerebral edema in acute liver failure. *Nat Clin Pract Gastroenterol Hepatol* 3: pp. 118–119, 2006.

[291] Obolenskaya MY, Vanin AF, Mordvintcev PI, Mulsch A, Decker K. EPR evidence of nitric oxide production by the regenerating rat liver. *Biochem Biophys Res Comm* 202: pp. 571–576, 1994.

[292] O'Grady JG. Acute liver failure. *Postgrad Med J* 81: pp. 148–154, 2005.

[293] Ohnhaus EE, Thorgeirsson SS, Davies DS, Breckenridge A. Changes in liver blood flow during enzyme induction. *Biochem Pharmacol* 20: pp. 2561–2570, 1971.

[294] Papasova M, Atanassova E. Adaption to surgical perturbations. In: *Handbook of Physiology—The Gastrointestinal System,* edited by Schultz SG, Wood JD, Rauner BB. Section 6, Volume 1, Chapter 33, New York: Oxford University Press, pp. 1199–1224, 1989.

[295] Parks DA. Ischemia-reperfusion injury: a radical review. *Hepatology* 8: pp. 680–682, 1988.

[296] Parks DA, Bulkley GB, Granger DN. Role of oxygen-derived free radicals in digestive tract diseases. *Surgery* 94: pp. 415–422, 1983.

[297] Parks DA, Bulkley GB, Granger DN. Role of oxygen free radicals in shock, ischemia, and organ preservation. Surgery 94: pp. 428–432, 1983.

[298] Parola M, Leonarduzzi G, Biasi F, Albano E, Biocca ME, Poli G, Dianzani MU. Vitamin E dietary supplementation protects against carbon tetrachloride-induced chronic liver damage and cirrhosis. *Hepatology* 16: pp. 1014–1021, 1992.

[299] Payen DM, Fratacci MD, Dupuy P, Gatecel C, Vigouroux C, Ozier Y, Houssin D, Chapuis Y. Portal and hepatic arterial blood flow measurements of human transplanted liver by implanted Doppler probes: interest for early complications and nutrition. *Surgery* 107: pp. 417–427, 1990.

[300] Pietroletti R, Chamuleau RAFM, Speranza V, Lygidakis NJ. Immunocytochemical study of the hepatic innervations in the rat after partial hepatectomy. *Histochem J* 19: pp. 327–332, 1987.

[301] Pinzani M, Failli P, Ruocco C, Casini A, Milani S, Baldi E, Gioitti A, Gentilini PJ. Fat-storing cells as liver-specific pericytes. Spatial dynamics of agonist-stimulated intracellular calcium transients. *Clin Invest* 90: pp. 642–646, 1992.

[302] Plaa GL. Toxicology of the liver. In: *Toxicology: The Basic Science of Poisons*, edited by Casarett LJ, Doull J. New York: MacMillan, pp. 170–189, 1975.

[303] Plaa GL, McGough EC, Blacker GJ, Fujimato JM. Effect of thioridazine and chlorpromazine on rat liver hemodynamics. *Am J Physiol* 199: pp. 793–796, 1960.

[304] Radi R, Beckman JS, Bush KM, Freeman BA. Peroxynitrite-induced membrane lipid peroxidation: the cytotoxic potential of superoxide and nitric oxide. *Arch Biochem Biophys* 288: pp. 481–487, 1991.

[305] Rappaport AM. The microcirculatory hepatic unit. *Microvasc Res* 6: pp. 212–228, 1973.

[306] Rappaport AM. Anatomic considerations. In: *Diseases of the Liver*, Fourth Edition, edited by Schiff L. Toronto: JB Lippincott, p. 36, 1975.

[307] Rappaport AM. In: *Liver and Biliary Tract Physiology I*, edited by Javitt NB. Baltimore: University Park Press, pp. 1–63, 1980.

[308] Rappaport AM. Microvascular methods—the transilluminated liver. In: *Hepatic Circulation in Health and Disease*, edited by Lautt WW. New York: Raven Press, pp. 1–13, 1981.

[309] Rappaport AM. The acinus—microvascular unit of the liver. In: *Hepatic Circulation in Health and Disease*, edited by Lautt WW. New York: Raven Press, pp. 175–192, 1981.

[310] Reilly FD, Dimlich RVW, Cilento EV, McCuskey RS. Hepatic microvascular regulatory mechanisms. II. Cholinergic mechanisms. *Hepatology* 2: pp. 230–235, 1982.

[311] Reilly FD, McCuskey RS, Cilento EV. Hepatic microvascular regulation mechanism. I. Adrenergic mechanisms. *Microvasc Res* 21: pp. 103–116, 1981.

[312] Reilly FD, McCuskey PA, McCuskey RS. Intrahepatic distribution of nerves in the rat. *Anat Rec* 191: pp. 55–67, 1978.

[313] Reilly WM, Saville VL, Burnstock G. Vessel reactivity and prejunctional modulatory changes in the portal vein of mature spontaneously hypertensive rats. *Eur J Pharmacol* 160: pp. 283–289, 1989.

[314] Richardson PDI, Withrington PG. Liver blood flow. I. Intrinsic and nervous control of liver blood flow. *Gastroenterology* 81: pp. 159–173, 1981.

[315] Richardson PDI, Withrington PG. Liver blood flow. II. Effects of drugs and hormones on liver blood flow. *Gastroenterology* 81: pp. 356–375, 1981.

[316] Richter S, Mucke I, Menger MD, Vollmar B. Impact of intrinsic blood flow regulation in cirrhosis: maintenance of hepatic arterial buffer response. *Am J Physiol Gastrointest Liver Physiol* 279: pp. G454–G462, 2000.

[317] Robard S. The burden of the resistance vessels. *Circ Res* 28 (Suppl 1): pp. 2–8, 1971.

[318] Rocheleau B, Ethier C, Houle R, Huet PM, Bilodeau M. Hepatic artery buffer response following left portal vein ligation: its role in liver tissue homeostasis. *Am J Physiol* 277: pp. G1000–G1007, 1999.

[319] Rockey DC. The cellular pathogenesis of portal hypertension: stellate cell contractility, endothelin, and nitric oxide. *Hepatology* 25: pp. 2–5, 1997.

[320] Rockey DC, Chung JJ. Inducible nitric oxide synthase in rat hepatic lipocytes and the effect of nitric oxide on lipocyte contractility. *J Clin Invest* 95: pp. 1199–1206, 1995.

[321] Rockey DC, Chung JJ. Regulation of inducible nitric oxide synthase in hepatic sinusoidal endothelial cells. *Am J Physiol* 271: pp. G260–G267, 1996.

[322] Rodriguez-Martinez M, Sawin LL, DiBona GF. Arterial and cardiopulmonary baroreflex control of renal nerve activity in cirrhosis. *Am J Physiol* 268: pp. R117–R129, 1995.

[323] Roland CR, Goss JA, Mangino MJ, Hafenichter D, Flye MW. Autoregulation by eicosanoids of human Kupffer cell secretory products: a study of interleukin-1, interleukin-6, tumor necrosis factor-alpha, transforming growth factor-beta and nitric oxide. *Ann Surg* 219: pp. 389–399, 1994.

[324] Rolando N, Wade J, Davalos M, Wendon J, Philpott-Howard J, Williams R. The systemic inflammatory response syndrome in acute liver failure. *Hepatology* 32: pp. 734–739, 2000.

[325] Rothe CF. Reflex control of veins and vascular capacitance. *Physiol Rev* 63: pp. 1281–1342, 1983.

[326] Rothe CF. Properties of veins. In: *Blood Vessels and Lymphatics in Organ Systems*, edited by Abramson DI, Dobrin PB. Orlando: Academic Press, pp. 85–96, 1984.

[327] Rothe CF, Stein PM, MacAnespie CL, Gaddis ML. Vascular capacitance responses to severe systemic hypercapnia and hypoxia in dogs. *Am J Physiol* 249: pp. H1061–H1069, 1985.

[328] Rubbo H, Radi R, Trujillo M, Telleri R, Kalyanaraman B, Barnes S, Kirk M, Freeman, BA. Nitric oxide regulation of superoxide and peroxynitrite-dependent lipid peroxidation. Formation of novel nitrogen-containing oxidized lipid derivatives. *J Biol Chem* 269: pp. 26066–26075, 1994.

[329] Sacerdoti D, Merkel C, Bolognesi M, Amodio P, Angeli P, Gatta A. Hepatic arterial resistance in cirrhosis with and without portal vein thrombosis: relationships with portal hemodynamics. *Gastroenterology* 108: pp. 1152–1158, 1995.

[330] Saftoiu A, Ciurea T, Gorunescu F. Hepatic arterial blood flow in large hepatocellular carcinoma with or without portal vein thrombosis: assessment by transcutaneous duplex Doppler sonography. *Eur J Gastroenterol Hepat*ol 14: pp. 167–176, 2002.

[331] Sancetta SM. Dynamic and neurogenic factors determining the hepatic arterial flow after portal occlusion. *Circ Res* 1: pp. 414–418, 1953.

[332] Sato N, Hayashi N, Kawano S, Kamada T, Abe H. Hepatic hemodynamics in patients with chronic hepatitis or cirrhosis as assessed by organ-reflectance spectrophotometry. *Gastroenterology* 84: pp. 611–616, 1983.

[333] Sato Y, Koyama S, Tsukada K, Hatakeyama K. Acute portal hypertension reflecting shear stress as a trigger of liver regeneration following partial hepatectomy. *Surg Today* 27: pp. 518–526, 1997.

[334] Sawchenko PE, Friedman MI. Sensory function of the liver—a review. *Am J Physiol* 236: pp. R5–R20, 1979.

[335] Schafer J, d'Almeida MS, Weisman H, Lautt WW. Hepatic blood volume responses and compliance in cats with long-term bile duct ligation. *Hepatology* 18: pp. 969–977, 1993.

[336] Schenker S, Bay M. Drug disposition and hepatoxicity in the elderly. *J Clin Gastroenterol* 18: pp. 232–237, 1994.

[337] Schoen JM, Wang HH, Minuk GY, Lautt WW. Shear stress-induced nitric oxide release triggers the liver regeneration cascade. *Nitric Oxide: Biology and Chemistry* 5: pp. 453–464, 2001.

[338] Schoen Smith JM, Lautt WW. The role of prostaglandins in triggering the liver regeneration cascade. *Nitric Oxide* 13: pp. 111–117, 2005.

[339] Segstro R, Greenway CV. α-Adrenoceptor subtype mediating sympathetic mobilization of blood from the hepatic venous system in anaesthetized cats. *J Pharmacol Exp Ther* 236: pp. 224–229, 1986.

[340] Segstro R, Seaman KL, Innes IR, Greenway CV. Effects of nifedpine on hepatic blood volume in cats: indirect venoconstriction and absence of inhibition of post synaptic α_2-adrenoceptor responses. *Can J Physiol Pharmacol* 64: pp. 615–620, 1986.

[341] Seyde WC, Longnecker DE. Anesthetic influences on regional hemodynamics in normal and hemorrhaged rats. *Anesthesiology* 61: pp. 686–698, 1984.

[342] Shen ES, Garry VF, Anders MW. Effect of hypoxia on carbon tetrachloride hepatotoxicity. *Biochem Pharmacol* 31: pp. 3787–3793, 1982.

[343] Shingu K, Eger II EI, Johnson BH, Van Dyke RA, Lurz FW, Cheng A. Effect of oxygen concentration, hypothermia, and choice of vendor on anesthetic-induced hepatic injury in rats. *Anesth Analg* 62: pp. 146–150, 1983.

[344] Siebert N, Cantre D, Eipel C, and Vollmar B. H_2S contributes to the hepatic arterial buffer response and mediates vasorelaxation of the hepatic artery via activation of K_{ATP} channels. *Am J Physiol Gastrointest Liver Physiol* 295: pp. G1266–G1273, 2008.

[345] Skandalakis JE, Gray SW, Soria RE, Sorg JL, Rowe Jr JS. Distribution of the vagus nerve to the stomach. *Am Surg* 46: pp. 130–139, 1980.

[346] Solis-Herruzo JA, Duran A, Favela V. Effect of lumbar sympathetic block on kidney function in cirrhotic patients with hepatorenal syndrome. *J Hepatology* 5: pp. 167–173, 1987.

[347] Souda K, Kawasaki T, Yoshimi T. Effects of acute and chronic ethanol administration on hepatic hemodynamics and hepatic oxygen consumption in awake rats. *J Hepatology* 24: pp. 101–108, 1996.

[348] Spitzer JA. Cytokine stimulation of nitric oxide formation and differential regulation in hepatocytes and nonparenchymal cells of endotoxemic rats. *Hepatology* 19: pp. 217–228, 1994.

[349] Spolarics Z, Spitzer JJ, Wang JF, Xie J, Kolls J, Greenberg S. Alcohol administration attenuates LPS-induced expression of inducible nitric oxide synthase in Kupffer and hepatic endothelial cells. *Biochem Biophys Res Comm* 197: pp. 606–611, 1993.

[350] Stadler J, Harbrecht BG, Di Silvio M, Curran RD, Jordan ML, Simmons RL, Billiar, TR. Endogenous nitric oxide inhibits the synthesis of cyclooxygenase products and interleukin-6 by rat Kupffer cells. *J Leukoc Biol* 53: pp. 165–172, 1993.

[351] Stark RD. Conductance or resistance. *Nature* 217: 779, 1968.

[352] Stipanuk MH. Metabolism of sulphur-containing amino acids. *Ann Rev Nutr* 6: pp. 179–209, 1986.

[353] Stramentinoli G, Gualano M, Ideo G. Protective role of *S*-adenosyl-L-methionine on liver injury induced by D-galactosamine in rats. *Biochem Pharmacol* 27: pp. 1431–1433, 1978.

[354] Suematsu M, Kumamoto Y, Sano T, Wakabayashi Y, Ishimura Y. Superoxide, NO and CO in liver microcirculation: physiology and pathophysiology. *J Hep Bil Pancr Surg* 3: pp. 154–160, 1996.

[355] Sutherland SD. An evaluation of cholinesterase techniques in the study of the intrinsic innervations of the liver. *J Anat* 98: pp. 321–326, 1964.

[356] Sutherland SD. The intrinsic innervations of the liver. *Rev Int Hepatol* 15: pp. 569–578, 1965.

[357] Takeuchi T, Horiuchi J, Terada N. Central vasomotor control of the rabbit portal vein. *Pflugers Arch* 413: pp. 348–353, 1989.

[358] Tandon HD, Tandon BN, Mattocks AR. An epidemic of veno-occlusive disease of the liver in Afghanistan. *Am J Gastroenterol* 70: pp. 607–613, 1978.

[359] Thames MC, Kinugawa T, Dibner-Dunlap ME. Reflex sympathoexcitation by cardiac sympathetic afferents during myocardial ischemia: role of adenosine. *Circulation* 87: pp. 1698–1704, 1993.

[360] Thews G. In: *Physiologie des Menschen*, 20th edition, edited by Schmidt RF, Thews G. Berlin: Springer, p. 543, 1980.

[361] Thurman RG, Kauffman FC, Jungermann K. *Regulation of Hepatic Metabolism*. New York: Plenum Press, 1986.

[362] Tran Thi TA, Haussinger D, Gyutko K, Decker K. Stimulation of prostaglandin release by Ca_2-mobilizing agents from the perfused rat liver. *Biol Chem Hoppe-Seyler* 369: pp. 65–68, 1979.

[363] Tyden G, Samnegard H, Thulin L. The effects of changes in the carotid sinus baroreceptors activity on splanchnic blood flow in anaesthetized man. *Acta Physiol Scand* 106: pp. 187–189, 1979.

[364] Um S, Nishida O, Tokubayashi M, Kimura F, Takimoto Y, Yoshioka H, Inque R, Kita T. Hemodynamic changes after ligation of a major branch of the portal vein in rats: comparison with rats with portal vein constriction. *Hepatology* 19: pp. 202–209, 1994.

[365] Ungvary G, Donath T. Neurohistochemical changes in the liver of guinea pigs following ligation of the common bile duct. *Exp Molec Pathol* 22: pp. 29–34, 1975.

[366] Urashima S, Tsutsumi M, Nakase K, Wang J-S, Takada A. Studies on capillarization of the hepatic sinusoids in alcoholic liver disease. *Alcohol Alcoholism* 28: pp. 77–84, 1993.

[367] Vandier C, Conway AF, Landauer RC, Kumar P. Presynaptic action of adenosine on a 4-aminopyridine-sensitive current in the rat carotid body. *J Physiol* 515: pp. 419–429, 1999.

[368] Van Leeuwen DJ, Sherlock S, Scheuer PJ, Dick R. Wedged hepatic venous pressure recording and venography for the assessment of pre-cirrhotic and cirrhotic liver disease. *Scand J Gastroenterol* 24: pp. 65–73, 1989.

[369] Villeneuve JP, Pomier G, Huet PM. Effect of ethanol on hepatic blood flow in unanesthetized dogs with chronic portal and hepatic vein catheterization. *Can J Physiol Pharmacol* 59: pp. 598–603, 1981.

[370] Wang JF, Greenberg SS, Spitzer JJ. Chronic alcohol administration stimulates nitric oxide formation in the rat liver with or without pretreatment by lipopolysaccharide. *Alcohol Clin Exp Res* 19: pp. 387–393, 1995.

[371] Wang H, Lautt WW. Does nitric oxide (NO) trigger liver regeneration? *Proc West Pharmacol Soc* 40: pp. 17–18, 1997.

[372] Wang H, Lautt WW. Hepatocyte primary culture bioassay—a simplified tool to assess the initiation of the liver regeneration cascade. *J Pharmacol Toxicol Meth* 38: pp. 141–150, 1997.

[373] Wang H, Lautt WW. Evidence of nitric oxide, a flow-dependent factor, being a trigger of liver regeneration in rats. *Can J Physiol Pharmacol* 76: pp. 1072–1079, 1998.

[374] Wang HH, McIntosh AR, Hasinoff BB, Rector ES, Ahmed N, Nance DM, Orr FW. B16 Melanoma cell arrest in the mouse liver induced nitric oxide release and sinusoidal cytotoxicity: a natural hepatic defense against metastasis. *Cancer Res* 60: pp. 5862–5869, 2000.

[375] Willet IR, Jennings G, Esler M, Dudley FJ. Sympathetic tone modulates portal venous pressure in alcoholic cirrhosis. *The Lancet* 11: pp. 939–943, 1986.

[376] Winwood PJ, Arthur MJP. Kupffer cells: their activation and role in animal models of liver injury and human liver disease. *Sem Liver Dis* 13: pp. 50–59, 1993.

[377] Wisse E. Ultrastructure and function of Kupffer cells and other sinusoidal cells in the liver. In: *Kupffer Cells and Other Liver Sinusoidal Cells*, edited by Wisse E, Knook DL. Amsterdam: Elsevier, pp. 36–60, 1977.

[378] Wisse E, Van Dierendonck JH, De Zanger RB, Fraser R, McCuskey RS. On the role of the liver endothelial filter in the transport of particulate fat (chylomicrons and their remnants) to parenchymal cells and the influence of certain hormones on the endothelial fenestrae. In: *Communications of Liver Cells*, edited by Popper H, Bianchi L, Gudat F, Reutter W. England: MTP Press Ltd, pp. 195–200, 1980.

[379] Wolfle D, Schmidt H, Jungermann K. Short-term modulation of glycogen metabolism, glycolysis and gluconeogenesis by physiological oxygen concentrations in hepatocyte cultures. *Eur J Biochem* 135: pp. 405–412, 1983.

[380] Woodhouse K. Drugs and the liver. Part III: Ageing of the liver and metabolism of drugs. *Biopharm Drug Dispos* 13: pp. 311–320, 1992.

[381] Yamaguchi N. Evidence supporting the existence of presynaptic α-adrenoceptors in the regulation of endogenous noradrenaline release upon hepatic sympathetic nerve stimulation in the dog liver in vivo. *Arch Pharmacol* 321: pp. 177–184, 1982.

[382] Yamaguchi N, Garceau D. Correlations between hemodynamic parameters of the liver and noradrenaline release upon hepatic nerve stimulation in the dog. *Can J Physiol Pharmacol* 58: pp. 1347–1355, 1980.

[383] Zhang JX, Pegoli WJ, Clemens MG. Endothelin-1 induces direct constriction of hepatic sinusoids. *Am J Physiol* 266: pp. G624–G632, 1994.

[384] Zink J. The fetal and neonatal hepatic circulation. In: *Hepatic Circulation in Health and Disease*, edited by Lautt WW. New York: Raven Press, pp. 227–248, 1981.

[385] Zink J, Greenway CV. Intraperitoneal pressure in formation and reabsorption of ascites in cats. *Am J Physiol* 233 (Heart Circ Physiol 2): pp. H185–H190, 1977.

[386] Zink J, Greenway CV. Control of ascites absorption in anesthetized cats: effects of intraperitoneal pressure, protein and furosemide diuresis. *Gastroenterology* 73: pp. 1119–1124, 1977.

[387] Zimmerman HJ. Chemical hepatic injury and its detection. In: *Toxicology of the Liver*, edited by Plaa G, Hewitt WR. New York: Raven Press, pp. 1–45, 1982.

[388] Zimmerman BJ, Granger DN. Oxygen free radicals and the gastrointestinal tract: role in ischemia-reperfusion injury. *Hepato-Gastroenterol* 41: pp. 337–342, 1994.

[389] Zimmon DS, Kessler RE. Effect of portal venous blood flow diversion on portal pressure. *J Clin Invest* 65: pp. 1388–1397, 1980.